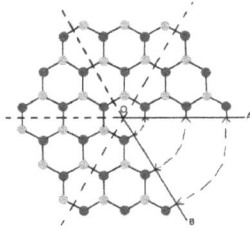

Many-Body Physics, Topology and Geometry

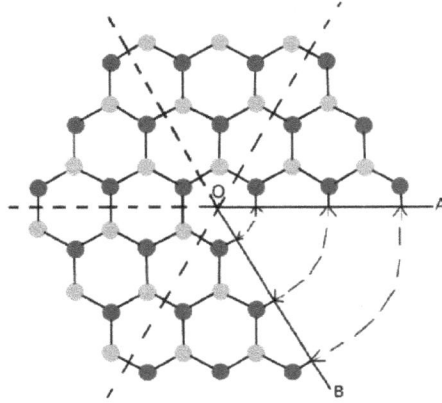

Many-Body Physics, Topology and Geometry

Siddhartha Sen
Trinity College Dublin, Ireland

Kumar Sankar Gupta
Saha Institute of Nuclear Physics, India

World Scientific

NEW JERSEY · LONDON · SINGAPORE · BEIJING · SHANGHAI · HONG KONG · TAIPEI · CHENNAI

Published by

World Scientific Publishing Co. Pte. Ltd.

5 Toh Tuck Link, Singapore 596224

USA office: 27 Warren Street, Suite 401-402, Hackensack, NJ 07601

UK office: 57 Shelton Street, Covent Garden, London WC2H 9HE

Library of Congress Cataloging-in-Publication Data
Sen, Siddhartha, author.
  Many-body physics, topology and geometry / Siddhartha Sen (Trinity College Dublin, Ireland),
Kumar Sankar Gupta (Saha Institute of Nuclear Physics, India).
    pages cm
  Includes bibliographical references and index.
  ISBN 978-9814678162 (hardcover : alk. paper) -- ISBN 9814678163 (hardcover : alk. paper)
  1. Many-body problem. 2. Topology. 3. Geometry, Differential. 4. Mathematical physics.
I. Gupta, Kumar Sankar, author. II. Title.
  QC174.17.P7S46 2015
  530.14'4--dc23
                                          2015013436

British Library Cataloguing-in-Publication Data
A catalogue record for this book is available from the British Library.

We dedicate this book to our teachers

Preface

In this book we explain a selection of ideas from mathematics and theoretical physics that are useful for understanding properties of physical systems of current interest such as graphene and topological insulators. The theoretical physics ideas discussed include an introduction to Quantum Field Theory which is a powerful qualitative tool in condensed matter physics that is well worth knowing. The mathematical ideas discussed are ones that have been used in condensed matter to explain remarkable results such as the quantisation of Hall conductance or the presence of gapless states. We also have a section illustrating the power of simple qualitative methods, which includes examples of observable consequences of zero point energy.

Quantum field theory is introduced in an intuitive way. The focus here is to quickly introduce the idea of a quasiparticle which encoded, in a single particle quantum framework, important features of a many body problem. We illustrate this feature through a number of examples. First for superfluids and superconductors and then, after giving a brief account of the Bogoliubov–de Gennes equations, for the more complex case of a topological insulator placed on a superconductor. Here a Majorana fermion quasiparticle can appear. In a separate section we give a rather more detailed account of the two-dimensional Dirac quasiparticle description of graphene. Using such a description, a number of situations are explored including the case of a graphene cone with a charge at its tip. The graphene cone problem is shown to be linked mathematically to a fundamental two-dimensional gravitational physics problem. Thus experiments with a graphene cone can, in principle, check theoretical gravitational physics results.

We explain mathematical ideas by presenting them in an intuitive way always stressing ideas and concepts. The presentation style is uneven. Some parts have a greater density of technical details than others but the aim is to give reasonably simple rules for applying the mathematical results presented to physical systems.

In the mathematics sections, we discuss differential geometry, topology and self-adjoint extensions. Topology is introduced as "rubber sheet" geometry and a discussion of the basic building blocks required to build such a geometry is followed by the introduction of the relevant concepts. The discussions on manifolds, differential forms, Stoke's theorem, cohomology, homotopy lead on to fibre bundles, K theory and Morse theory. Physical examples where these ideas are used are discussed with special attention given to the fact that topological reasoning can predict the existence of gap less points.

The method of self-adjoint extensions is discussed through examples which illustrate their need and usefulness for graphene and other physical systems such as the polar molecules. In these examples the method of self-adjoint extensions is used to introduce short distance features in an effective long distance model through boundary conditions.

We have tried to keep the prerequisites for following arguments presented to a minimum. What is required is interest and the ability to try to understand and follow unfamiliar arguments.

The book should be of interest to students of physics and all curious students of nature.

KSG would like to thank A. P. Balachandran for introducing him to the subject of self-adjoint extensions and for many inspiring discussions. KSG would also like to thank B. Basu-Mallick, Baishali Chakraborty, Sayan K. Chakrabarti, Pulak R. Giri, Pijush K. Ghosh, Stjepan Meljanac, Andjelo Samsarov and S. G. Rajeev for collaboration and discussions on some of the materials included in Chapters 4 and 5 of this book. Many of the plots and some of the materials in Chapters 4 and 5 are taken from papers written jointly with some of them. KSG would also like to thank Baishali Chakraborty for help with some of the diagrams in Chapter 5. SS would like to thank Koushik Ray and Krishnendu Sengupta for discussions and M. Coey both for wide ranging discussions and for initiating this project.

Siddhartha Sen and Kumar S. Gupta

Contents

Chapter 1

Overview

The challenge of condensed matter physics is to use non relativistic quantum ideas to explain and predict the observed macroscopic properties of matter. To do this great ingenuity and imagination is required. The Hamiltonian H of a many-body system can be written down schematically as

$$H = H_0 + H_I$$

where H_0 represents the sum of free Hamiltonians representing the constituent particles (atoms/molecules) of the system, while H_I represents the interactions between these particles. These interactions are usually represented by static potential energy terms where the potential energy is taken to be the sum of two-body potentials i.e. potentials which only depend on the positions of pairs of particles. Many-body potentials, where the potential energy depends on the coordinates of more than two particles, are usually not considered, but seem to be required to understand certain systems, for example for modeling water. They also sometimes appear from theoretical calculations. For example the Casimir potential between a collection of parallel strings depends on all the string positions.

Determining the ground state and excited state of a complex system ab initio starting from an arbitrary collection of atoms and molecules by analytic or numerical means is an impossible task even with the simplification of considering only two-body potentials. There are just too many particles and too many possible interactions. But the outlook for the subject is far from being bleak. Special models and simplifying assumptions, motivated by physical insights, have very successfully explained the basic nature of insulators, transistors and conductors. Models and general arguments have also successfully explained the nature of normal phase transitions such as in magnetism or s-wave superconductivity and superfluidity. Theoretical models have clarified many conceptual problems such as the possibility of fractional charge, the possibility of solitonic excitations and even the possibility of non bose/fermi statistics.

In all these successful models the nature of the ground state is taken to be given i.e. the ground state is either a solid or a liquid or a gas. The constituents of the system are also assumed and the arrangement of the constituents is also assumed to be known, for instance for a crystal. Even these assumptions are not enough to tackle the problem and further approximations are needed to determine the properties of the system.

There is also a conceptual problem we would like to highlight. The assumption that the properties of a complex system can be determined from a knowledge of its constituent free particles and the interaction between them might not be correct. There could be new emergent quantum particle excitations that are essential for understanding the system. An example of this possibility is a crystal lattice where besides the atoms, the emergent phonon excitation that represents lattice vibrations, has to be included. This excitation only appears when a crystal is formed. The crystal is thus a macroscopic stable quantum ground state with the phonon as the emergent quantum excitations which characterizes the system. The phonon only appears when the crystal is formed. It is not present in the starting description of the system.

There are also new challenges emerging that come from the rapid growth of technology and there are new theory driven areas of exploration. Technological growth has led to the fabrication of new materials with novel properties so that it is now possible to engineer atoms and molecules and even light beams to form new structures at the nanoscale.

These technology driven challenges continue to show that the properties of matter at the nano scale are full of surprises. Understanding them require new insights. These insights can come from experimental clues, from physical insights or from an imaginative use of mathematical ideas.

Thus using ideas of topology the topological insulator was predicted to be possible and subsequently discovered. Similarly, topological considerations lead to the prediction of Majorana fermions in condensed matter systems. Another powerful theoretical idea is the suggestion that a different type of phase transition called a quantum phase transition is possible in which the ground state of a system changes as the coupling constants of the system change. Unlike a normal phase transition where there is an order parameter, such as magnetization, which is zero in one phase and non zero in the other phase and is temperature dependent, the new type of phase transition has no order parameter. It happens at zero temperature.

It is property of the ground state. A proper universally agreed description of such a transition, with examples is not available. But what is becoming clear is that such transitions are present and that topological ideas are likely to play a role in understanding some of them.

The imaginative use of mathematical insights we are talking about do not involve new methods of solving or making good approximations but are new ways of thinking about problems.

An example of such a mathematical idea is the method of self-adjoint extensions. This method helps us to obtain appropriate boundary conditions in a way that preserves the basic physical requirement of unitarity. This is very useful in dealing with singular interactions, where the boundary conditions play a very important role in obtaining the correct physical results.

Another class of useful mathematical ideas come from differential geometry, topology and group theory. These methods help us to identify, for example, systems that can have a zero gap in their energy spectrum or when response functions of a system, such as its conductance, are quantized as in the quantum Hall effect. The methods of topology reveal physical properties that depend only on global features of wave functions of the system. They are powerful ways of extracting information about complex systems. We will illustrate how these ideas work by looking at examples such as graphene and topological insulators. In order to present topological ideas in a suitably general setting we have included a discussion of differential manifolds and differential geometry which introduce concepts that are necessary for discussing the idea of a fibre bundle. The abstract language of fibre bundles may not be familiar but once an intuitive picture of the underlying geometrical picture used is clearly understood the power of the abstract language to tackle problems of physical importance should become apparent. The language of fibre bundles is useful for understanding the topological insulator. The way we present mathematical ideas is to explain why they are important for physics, then describe what they are in an intuitive way, followed by explaining how the mathematical ideas introduced can be used to carry out calculations.

The backdrop in which these mathematical ideas are presented is quantum mechanics and quantum field theory. For this reason we also include a brief intuitive account of quantum field theory with a focus on two topics.

The first topic we will consider is the quasiparticle. Quasiparticles are single particle excitations that are constructed to capture the essential physics of a complex system within a specified energy/momentum range.

The idea has been successfully used to understand a variety of systems such as superfluids, superconductors, graphene and topological insulator. We will consider these examples.

The second topic which we will discuss briefly is less well established but is gaining attention in the world of nanophysics and nanobiology. It is the idea that nanoscale coherent structure can form spontaneously. This can be shown by considering charged particles, described by a quantum field theory, interacting with a time- dependent vector potential. The vector potential represents photons in quantum field theory. A surprising result of the quantum uncertainty principle is that even when no photons are present the ground state still has energy. It has energy called zero-point energy and is due to vacuum fluctuations required by the uncertainty principle. We will see that vacuum fluctuations are responsible for the decay of atoms from stationary excited states to the ground state. We will also show how such zero-point interactions can lead to spontaneous coherent nanoscale structures for an assembly of charged particle when their density crosses a certain theoretically determined critical value. The presence of these structures will have experimental consequences. The emergence of such structures provides an example of a new dynamic nanoscale phase.

Let us end our general discussion by saying that topological features show up in physics through boundary conditions. Very often physical systems are described by differential equations. The same equation $\nabla^2 \phi = \rho$ shows up in electrostatics and in gravity. The difference between systems lie in the boundary conditions used. If, for instance, we would like to calculate the electric potential at a point on the torus when there are no charges present then we would need to solve $\nabla^2 \phi = 0$ with boundary conditions that capture the geometry of the torus. But the torus also has topological properties and these properties can be used to draw general conclusions regarding the nature of this potential. We will discuss the topological properties of the torus later.

We now give a few examples of topological ideas.

Topological Ideas: Euler's Genus Formula

We start by writing down Euler's famous topological result for a a closed surface constructed by gluing together different flat N-gons (N sided surfaces such as triangles, squares, pentagons, heptagons etc). Such a way of constructing a closed surface is called a triangulation. Euler's result states that

$$V - E + F = 2 - 2g$$

where g is the genus of the surface, V, E, F are the number of vertices, edges and faces present in the triangulation. This is a truly remarkable result. There are many different ways of triangulating a given surface. For example a sphere can be triangulated by four triangles to form a tetrahedron or by six squares to form a cube. For both these triangulations $V - E + F = 2$. Thus different ways of triangulating a surface will have different values for V, E and F but the great discovery of Euler was that the combination $V - E + F$ did not depend on the shape or scale of the surface but was the same for surfaces which had the same connectedness property . Thus all sphere-like shapes have $V - E + F = 2$ (genus zero), all tori-like shapes have $V - E + f = 0$ (genus one) and so on. The genus represents the number of holes present in the surface.

Lower dimensional version of Euler's result also hold. Thus if a line is made by one edge with two vertices at its ends we have $V - E = 1$. If a longer line is formed by joining together two such units it will have a different length but its $V - E = 1$, remains unchanged as it now has three vertices and two edges. The two lines have different lengths and are geometrically different but their $V - E$ is the same. For this reason the invariant of Euler is called a topological invariant.

We can do more. If three edges are used to form a triangle or four edges are used to form a square, or five edges are used to form a pentagon we can check that $V_E = 0$ for all these cases. Indeed for an arbitrary network we have the result $V - E = -(N - 1)$ which tells us that the network has N independent loops.

A higher dimensional version of Euler's result exists and was established by Poincare using his topological invention of the Homology group.

Let us establish another simple but important consequence of this result which is also due to Euler. Suppose we want to construct a genus g surface using different N-gons glued together (an N-gon is an N sided surface) to form the a genus g closed surface. In order to form a closed surface all the edges of the constituent N-gons must be glued so that after gluing no free edge remains. We also assume, for simplicity, that in joining pieces together three vertices are always to be glued together. We can now proceed. First we note

(1) A N-gon has N vertices
(2) A N-gon has N edges

(3) Each N-gon has one face

Thus for a collection of different N-gons we have:

(1) The number of N sided faces present in this collection is, we suppose, F_N
(2) Then the total number of edges for this collection of different N-gons is: $\sum N F_N$
(3) Then the total number of vertices for this collection of different N-gons is $N F_N$

Using our rules for gluing two edges to eliminate boundaries and gluing three vertices we get the vertices V_T, edges E_T and faces F_T of the triangulation as:

(1) The total number of faces $F_T = \sum F_N$
(2) The total number of edges $E_T = \frac{1}{2} \sum N F_N$
(3) The total number of vertices $V_T = \frac{1}{3} \sum N F_N$

We now use Euler's result and set

$$
\begin{aligned}
2 - 2g &= V_T - E_T + F_T \\
&= \frac{1}{6} \sum (6 - N) F_N
\end{aligned}
$$

This is Euler's result. Note that N-gons with $N = 2, 3, 4, 5$ give a positive contribution to the sum, $N = 6$ gives zero contribution while for $N \geq 7$ the contributions are negative. Thus to construct surfaces for which $2 - 2g \leq 0$ N-gons with $N \geq 7$ will be required. Adding $N = 6$ sided faces will not change the genus. In order to get a $g = 0$ surface we can simply choose $N = 5$ then Euler's formula tells us that

$$
2 = \frac{1}{6} F_5
$$

which gives $F_5 = 12$ i.e. twelve pentagons joined together form a genus zero, surface, which is a sphere-like. This used to be the way a football (soccer ball) was made.

We can draw an important conclusion from this result. The sphere has genus zero and is a surface which has positive curvature. The torus, genus one, can be constructed using only hexagons. It has zero curvature, while all higher genus surfaces have negative curvature. Thus there seems to be a link between curvature and the number of sides of an N-gon. The link between N-gons and curvature can be made explicit using the following model. Start with the $N = 6$ hexagon. It has zero curvature. It is flat.

We can think of the hexagon as being made out of six equilateral triangles. From this basic result we construct a pentagon by removing one of the six equilateral triangles of a hexagon and then gluing the sides exposed by the removal of the wedge together. This forces the pentagon to curve up with positive curvature. Similarly if an additional equilateral triangular wedge is introduced in the hexagon to create a 7-gon then negative curvature is introduced. This rigid geometric model for the N-gons establishes a link between an N-gon and curvature.

Topological Example: Fullerene Production in an Arc

We now use these ideas to understand what percentage of carbon atoms in an arc form fullerene. We recall that in 1985 Kroto and collaborators discovered fullerene molecule which they showed was composed of 60 carbon atoms, with 12 pentagons and 20 hexagonal rings. From our discussion of Euler's result we immediately know that this molecule is a triangulation of a sphere-like surface, with each pentagon carrying the same amount of positive curvature while the hexagons present have zero curvature. Fullerene is formed in great abundance. They are present, for example, at the level of $\approx 10\%$ of the 25% of Carbon atoms present in the electric discharge between two graphite electrodes.

Can this result be understood using a simple geometrical model? Topology already tells us that the molecule is a closed structure representing a tiling of a sphere-like shape. We now need some geometry.

The geometrical constraints gives a simple rule which tells us how pentagonal and hexagonal rings of carbon should be assembled in order to form fullerene. We have the rules: two pentagons can never share an edge nor can three hexagons share a common vertex as these configurations prevent the formation of fullerene. The basic building block is two hexagons and one pentagon which is C_{12}. This molecule thus has three rings. Two hexagonal rings and one pentagonal ring. Fullerene has 32 rings. Thus starting from C_{12} we need to add 23 more rings to C_{12} to get fullerene. If the probability for a ring to form is q then the percentage of fullerene expected starting with 25% carbon is

$$25\% q^{23}$$

To get an yield of 10% means $q \approx .961$. Let us sketch how q can be determined. In general we have the following possibilities of adding pentagon and hexagonal rings. Each possibility is assigned a Boltzmann factor. Thus we have the following four possibilities:

(1) A pentagon created between two hexagons: Boltzmann factor $e^{-\epsilon}$.
(2) A pentagon created between a pentagon and a hexagon: Boltzmann factor $e^{-\beta}$.
(3) A hexagon created between two hexagons: Boltzmann factor $e^{-\alpha}$.
(4) A hexagon created between a pentagon and a hexagon: Boltzmann factor $e^{-\eta}$.

The Boltzmann factors represent the energy of a configuration. Using these expressions we can calculate the normalised probability of producing clusters that can lead to the formation of fullerene. For example when the four-polygon molecules form with only two that can lead to fullerene formation their total probability is

$$F_4 = \frac{e^{-\epsilon} + 2e^{-\eta}}{e^{-\epsilon} + 2e^{-\eta} + e^{-\alpha} + 2e^{-\beta}}.$$

Thus if α, β, the disallowed configurations are large we can easily understand the large percentage of fullerene observed. Kerner has carried out such an analysis and was able to get a yield value of fullerene of about 9.5% without assuming values for α, β but used an additional hypothesis of self similarity during the process of assembly. The purpose of this example is to show how an unusual combination of topology and geometry can be used to understand an experimental result and also to point out how ideas of probability can be combined with geometrical ideas. The simple expectation regarding disallowed configurations could, in principle, be checked by molecular calculations.

Chapter 2

Many-Body Theory

2.1 Introduction

Condensed matter physics deals with systems that have a large number of atoms and molecules close together on the atomic scale. For this reason the appropriate theoretical framework for understanding condensed matter physics is quantum theory, which was created in order to explain atomic scale phenomena. In fact, as we will see, the appropriate framework for many situations, is not quantum mechanics but quantum field theory. Quantum mechanics deals with system which have a finite number degrees of freedom while quantum field theory deals with systems that have an infinite number of degrees of freedom.

Condensed matter physics uses insights and imagination to replace complex systems by simple models that capture their essential features. These simple models work remarkably well. Thus reasonable questions to ask are: What are useful ideas for model building and what are useful ways for gaining physical insight?

We intend to provide practical answers to these questions. Useful ideas for model building are identified by examining common features present in successful models and then two ways for gaining physical insight are suggested. The first way is by using qualitative arguments. In this approach the essential features of a problem are identified on the basis of experimental information. Thus for molecules that form a non-conducting lattice it is reasonable to assume that the electrons of the molecules are confined, that is, they are tightly bound. The next step is to build a quantum field theory model incorporating this information, work out observable consequences that follow from the model and then make measurements to check if they are correct. We illustrate this approach by a number of examples.

The second way of gaining insight is by using mathematical ideas. The power of this approach is illustrated in two separate chapters in which two sets of mathematical ideas are discussed. The first chapter is on topology and the second on the method self adjoint extensions. The methods of Topology can be used to explain the existence of gap less states or the quantisation of Hall conductance. Topological results of this kind follow from very general considerations and are not influenced by the detailed nature of the interactions present, while the method of self adjoint extensions can be used to incorporate short distance information in an effective long distance theory.

After introducing the basic ideas of quantum field theory (QFT) we use QFT to model a number of systems such as graphene, superconductors, superfluids and topological insulators. The examples chosen are ones where "quasiparticles" play an important role. A quasiparticle is an emergent effective single particle excitation, valid in a certain energy range, associated with a many-body system. The quasiparticle can be used to study how the system responds to external disturbances, using the methods of quantum mechanics rather than those of QFT. If, however, we are interested in physical properties which depend on the interaction between quasiparticles then a QFT approach is necessary. For example electrons in graphene are replaced by two dimensional mass-less Dirac quasiparticles, valid for low energies, but if we want to study conductance properties of graphene we need to calculate the scattering cross section between these quasiparticles. To do this QFT methods have to be used.

Let us start our journey of explaining theoretical tools and concepts of condensed matter physics with Bohr's model for hydrogen. The Bohr atom illustrates the power of a simple model to capture the essential physics of a system.

After discussing Bohr's atom we move on to further illustrate the power of qualitative reasoning by discussing different examples including consequences of the fact that photons have zero point energy. This requirement of quantum theory tells us that even when no photons are present there is still energy. This astonishing consequence of quantum theory has many verified experimental consequences. For instance it explains energy changes in the spectrum of hydrogen, the Casimir effect and why higher excited states of isolated atoms have lifetimes.

Let us begin. We start with the way Bohr tried to understand the observed regularities of spectrum of hydrogen.

2.1.1 *The Spectrum of Hydrogen*

In the months of July, September and November in 1913 Niels Bohr submitted three paper that proposed a revolutionary theory for understanding the light emitted by hydrogen atom. It used Newton's ideas of mechanics but rejected Maxwell's ideas regarding radiation emitted by an accelerated charged particle. This theory is now taught as one of the defining steps that led to quantum mechanics. It introduced the new idea that light was emitted only when an electron jumps from one allowed energy level to another. The frequency of the emitted light was not directly related to the motion of the electron in an allowed energy level. The light magically appears from nowhere to save the conservation of energy law.

It is always pleasant to recall well known arguments that have profoundly changed our views regarding physics. Bohr's argument for hydrogen is certainly such a heroic step and a turning point for physics. It was Bohr's habit before tackling a conceptually challenging problem to retrace the struggle that the physics community faced in order to resolve an older problem. Such a retelling of previous triumphs prepared the mind to face new challenges. We are now facing challenges in understanding the composition of the universe, for understanding complex systems such as why proteins fold, how the brain works, the mysteries of the nanoworld, and for constructing a grand unified theory of all the forces of nature. It is thus a good time, following Bohr's lead, to revisit the triumph of understanding the spectrum of hydrogen proposed a hundred years ago and to explain the novelty and daring of his model. Let us review the basic features of the model. They can be listed as follows:

(1) Rutherford had studied the way highly energetic charged particles scattered from hydrogen. His experimental findings showed that most of the time the charged particles went straight through but occasionally they were hurled back: they suffered large angle scattering. Rutherford found that he could explain these results by a simple solar system like model for hydrogen with the proton at the centre and the electron orbiting round it. This simple model replaced an earlier plum pudding model for atoms due to Thomson.

(2) However Rutherford's model was unstable on theoretical grounds. This was because Maxwell's laws of radiation required a circulating charged particle to radiate light and thus lose energy and ultimately to collapse on to the proton in a fraction of a second. Furthermore the light emitted by the circulating and spiraling electron would have frequencies spread

over a continuous range reflecting the changing orbital frequency of the electron as it collapsed.

(3) On the other hand hydrogen atom was experimentally observed to be stable, have a size much bigger than that of the proton and the light it emitted had precise discrete frequencies ν_n extremely well described by the empirical formula found by Balmer in 1885, namely

$$\nu_n = R(\frac{1}{2^2} - \frac{1}{n^2})$$

In order to explain what was observed Bohr boldly supposed that

(1) No radiation occurred from some special discrete set of orbits which he called stationary orbits. These special orbits had angular momentum l which were integral multiple of Planck's constant h. Thus unlike the solar system where orbits of say, satellites, depend on initial conditions these special stationary orbits were fixed by nature. They did not dependent on initial conditions. This was a profound idea.

(2) Radiation of light occurred when an electron made a jump from one discrete stationary state of energy, E_m to another, of energy E_n. The frequency of light emitted was postulated to be given by

$$h\nu_{nm} = E_n - E_m$$

The light emerged miraculously when the electron made such a transition. Again a very revolutionary idea.

Thus a quantum idea of discreteness was used twice. Once to identify stationary orbits and their energies and again to relate energy differences to the frequency of the emitted light. The idea that energy of light appeared as quanta had been suggested by Planck and Einstein. But the idea that angular momentum had a quantum nature was a new idea.

Simple mathematical steps from this postulates led Bohr to Balmer's formula and a prediction for the constant R. Let us go through these steps by assuming that the electron of mass m and charge $-e$ revolves in a stationary circular orbit of radius r, with speed v. Then the angular momentum of the electron, fixed by Bohr's condition is

$$mvr = l = nh$$

On the other hand Newton's law requires

$$m\frac{v^2}{r} = \frac{e^2}{r^2}$$

where we have used Coulomb's law of attraction between the electron and proton and the fact that a circular motion of speed v has acceleration $\frac{v^2}{r}$. We also know that the energy E_n of the electron is

$$E = \frac{1}{2}mv^2 - \frac{e^2}{r}$$

Using Newton's law we can write the energy expression as

$$E_n = -\frac{1}{2}\frac{e^2}{r}$$

On the other hand combing the stationary orbit condition with Newton's law gives

$$mv^2 = \frac{n^2 h^2}{mr^2} = \frac{e^2}{r}$$

Thus

$$\frac{1}{r} = \frac{me^2}{n^2 h^2}$$

So that we finally get

$$E_n = -\frac{me^4}{2n^2 h^2} = -R\frac{1}{n^2}$$

We have stressed the revolutionary nature of Bohr's model with its a la carte approach to electromagnetic theory. Thus Coulomb's law was used but Maxwell's law of radiation was disregarded in tackling a problem that had to do with the radiation and emission of light! How could this happen? How could one propose a theory of light radiation without a theory of light emission present? Just stating that light miraculously appeared during a jump from one level to another was not satisfactory. Bohr was completely aware of these difficulties but his intuition told him that understanding the light emitted by hydrogen required revolutionary ideas. It could not be understood by simple modifications of existing theories.

The next step in understanding the spectrum of hydrogen came with the invention of quantum mechanics by Heisenberg. The idea of Heisenberg was to build a theory at the atomic scale following Einstein's idea of focusing on observables. Thus the analysis of Bohr suggested to Heisenberg that hydrogen could be regarded as a collection of virtual oscillators with Bohr frequencies $\nu_{nm} = \frac{E_n - E_m}{h}$. Let us develop this idea in a simple minded way. We want to discover the hidden dynamics contained in this formula. Thus we set

$$x_{mn}(t) = e^{-i\nu_{nm}}$$

Differentiating this and using Bohr's frequency condition gives

$$i\frac{dx_{mn}(t)}{dt} = \nu_{mn}x_{mn}(t)$$

$$= \frac{E_n - E_m}{h}x_{mn}$$

This equation Heisenberg interpreted, with the help of Born, as a matrix equation. Later Born learned from Weiner that such equations could be understood as equations involving linear operators. Thus replacing the matrices $x_{mn}(t)$ by the linear operator $x(t)$ and regarding the energies E_n, E_m as energy eigenvalues of a linear energy operator H we get

$$ih\frac{dx(t)}{dt} = [H, x(t)]$$

where $[H, x(t)] = Hx(t) - x(t)H$ is the commutator. Setting H to be the Hamiltonian then gives an equation of motion for the linear operator $x(t)$. The commutation relations between the linear operators representing the position and the momentum follow as consistency requirements i.e. from the requirement that

$$ih\frac{dx(t)}{dt} = [H, x(t)]$$

$$H = \frac{p^2}{2m} + V(x)$$

For these to form a consistent set of algebraic equations we must have

$$ih\frac{dx(t)}{dt} = \frac{1}{2m}(p[p, x] + [p, x]p)$$

$$= \frac{p}{m}[p, x]$$

from which it follows that $[p, x] = ih$ as $\frac{p}{m} = \frac{dx}{dt}$. We assume that $[V(x), x] = 0$. Generalising the observation that the commutator of the linear operator $x(t)$ with H gives its time evolution to all linear operator gives the dynamics of this new mechanics. This method was used by Pauli to work out the spectrum of hydrogen using a special symmetry present for a $\frac{1}{r}$ and quantum mechanics was created. In Pauli's approach there was no need to postulate stationary states and Bohr's result was obtained. Soon Schroedinger formulated Wave Mechanics and Dirac used Poisson brackets to give another formulation of quantum mechanics followed later by Feynman's path integral formulation. In the early hectic period of working out consequences of the new theory Heisenberg stepped back from the excitement of explaining hitherto mysterious results of atomic physics and discovered a profound conceptual feature of quantum theory called the Uncertainty Principle. This Principle, he showed, followed from the fact that operators, in quantum theory, that represent physical observables need not commute. For a pair of such operators the product of the uncertainties of the operators on a given state can be non zero. This is the Uncertainty Principle. More precisely if the uncertainty $\Delta A(s)$ of an operator A in a given normalised state $|s>$ is defined to be $\Delta A(s) = \sqrt{<s|(A-<s\,A|s>)^2|s>}$ then the Uncertainty Principle for a pair of operators A, B on a given state $|s>$ is the statement that

$$\Delta A(s)\Delta B(s) \geq \frac{1}{2}|<s|[A,B]|s>|$$

where $[A, B] = AB - BA$ is the commutator. For a harmonic oscillator this result applied to the operators x, p implies that the lowest energy state for a quantum oscillator, corresponding to a classical oscillator with classical oscillation frequency ω, is not zero but $\frac{1}{2}\hbar\omega$. The zero point energy is a purely quantum phenomenon. Higher excited states of the oscillator have quantum energies of the form $(n + \frac{1}{2})\hbar\omega$, where n is an integer. Using a generalisation of the Heisenberg Uncertainty Principle due to Sobolev the stability of hydrogen atom in quantum theory and, much latter, the stability of matter established by Lieb and Thirring. A hurdle in our understanding of nature had been crossed.

In Bohr's model, and even in quantum mechanics models of hydrogen, the fact that photons emerged in a transition was simply accepted as a miracle. There was no photons in the theory. Thus to properly understand the spectrum of hydrogen a quantum version of Maxwell's theory of photons interacting with electrons was required. Such a theory, quantum electrodynamics, was eventually constructed and has proved to give a stunningly accurate description of atomic phenomena. Using this theory the spectrum of hydrogen was finally properly explained.

For later use we now briefly comment on the quantum description free photons. Photons are represented by a vector potential $A(x, t)$ which satisfies the vector potential equation,

$$\nabla^2 A(x, t) - \frac{\partial^2 A(x, t)}{c^2 \partial^2 t} = 0$$

where the vector nature of $A(x, t)$ has been suppressed. For interacting photons the right hand side will be a non zero current. Solving this classical equation in, say a cubic enclosure, gives the vector potential as a sum of oscillating terms with frequencies fixed by the cubic box size. Each oscillating term can be regarded as a classical harmonic oscillator which means that we can, following this line of thought, construct a quantum description of free photons in terms of harmonic oscillators. We will show later, that the higher excited state label n of a quantum harmonic oscillator has to be interpreted in such a quantum theory for photons, as the number of photons present. But each oscillating term, regarded as a classical oscillator, will have zero point energy, which corresponds to the energy with $n = 0$ i.e. when no photons are present. Thus even in the absence of photons there is photonic energy present in a quantum theory of photons. For later use we also recall that the electric field $E(x, t)$ is related to the vector potential by as follows,

$$E_i(x, t) = \frac{1}{c} \frac{\partial}{\partial t} A_i(x, t) + \nabla_i A_0(x, t)$$

where c is the velocity of light in free spaced, and $A_0(x, t)$ is the scalar potential.

2.1.2   *The Power of Qualitative Reasoning*

One of the joys of physics is to estimate the order of magnitude of the quantities of interest and to try to predict their behavior using simple ideas. Such an approach was, for instance, used by Gamow to estimate the temperature of the universe.

There are two useful ways of making estimates. The first way is to use dimensional analysis. In this approach we need to decide on the variables the observable of interest depend on. Once this is done the next step is to form the dimension of the observable of interest out of variables chosen. In general this will lead to a functional relationship between physical variables. Using such an approach the period of oscillation of a pendulum or the period of revolution of a planet round the Sun can be found.

Thus for the pendulum if we suppose the oscillation time T is related to the rigid string length L and the gravitational attraction g then matching dimensions we see that T and $\sqrt{\frac{L}{g}}$ have the same dimension so we set $T = k\sqrt{\frac{L}{g}}$, where k is a dimensionless constant.

While for the rotation period T if we suppose it depends on the orbital radius R, the mass of the Sun M and the gravitational constant G, again matching dimensions we see that T and $R^{\frac{3}{2}} \frac{1}{\sqrt{GM}}$ have the same dimension so can set $T = kR^{\frac{3}{2}} \frac{1}{\sqrt{GM}}$, with k a dimensionless constant.

The second way is to build a simple models for the system and then analyse it in a qualitative way using the laws of physics. This approach needs greater insight. Thus for both the pendulum and the revolution time period of a planet can both be understood from the gravitational law of attraction and the force law of Newton. However the qualitative approach does not use the dynamical laws in detail. Very often the important conservation laws of energy and momentum that follow from the laws are enough to make qualitative estimates.

Let us now illustrate the method of qualitative reasoning by giving a few examples.

2.1.3 *Estimate of Speed of Electrons and Size of Their Orbits*

We want to estimate the magnitude of an inner electron's velocity in an atom orbit as well as the orbit size. This is done by constructing a quantity with the dimensions of velocity from the variables that describe the structure of the atom. Listing these variables thus requires physical insight. A reason for including each variable on this list has to found. Thus

(1) For atomic scale phenomenon we expect quantum theory to be relevant thus we must include $\hbar$
(2) Orbits form due to Coulomb charges of the nucleus thus the charge Ze must be included.
(3) The motion of the electron will depend on its mass m and its charge e. These must be included
(4) If we assume the atomic system is non relativistic then, c the velocity of light should not be on the list. Similarly gravitational interactions are small compared to Coulomb interactions and are thus not in the list.

Dimensional analysis immediately tells that there is only one combination of $Z, e, m, \hbar$ that has the dimensions of velocity it is

$$\text{velocity} \rightarrow \frac{Ze^2}{\hbar}$$

Note there is no m dependence. To estimate the size of the orbit we use a simple but powerful idea for qualitative estimates namely that for a system in dynamic equilibrium its kinetic energy and potential energy are proportional. This is a "virial theorem" which can be easily proved for a circular electron orbit under a coulomb force but we skip the proof. Assuming a

circular orbit of size a and setting the kinetic energy of an electron equal
to its potential energy we get,

$$\frac{1}{2}mv^2 = \frac{Ze^2}{a}$$

which gives $a = \frac{\hbar^2}{Zme^2}$. In the approach the linear dependence on the
number Z follows from the physics of the Coulomb interaction between the
electron of charge e and the nucleus of charge Z. Thus each of the choices
made can be justified by a physical argument and we expect our result to
be reasonable.

2.1.4   *Charge Distribution in an Atom*

We want to determine the way the electrons are distributed in a heavy atom
and to estimate the radius R of the region containing most of them. The
number of electrons in such an atom will be high. We now make a dras-
tic simplifying assumption and then suggest that it is reasonable because
electrons are fermions. Our assumption is that in the region with most
electrons the potential is constant and that the electrons can be treated as
free particles. These assumptions can be justified using the localised nature of wavefunctions and the fact that electrons are fermions. For densely
fermions packed all quantum states are filled hence interactions between
electrons cannot change their state thus, it is then surprising but reasonable, to such treat such electrons as free particles. Except for orbits very
close to the nucleus the variation of the potential is small, falling off as a
at least a quadratic power of distance. Hence in a the localised region of
maximum electron numbers the spatial dependence of the potential can be
neglected. For free electrons, the charge density, given by the number of
electrons per unit volume is

$$n = 2 \int \frac{d^3p}{2\pi^3} = \frac{p_0^3}{3\pi^2}$$

where the factor of two reflects the two spin states and p_0 is the maximum
momentum value of electrons that fill the orbits can have. We can relate
this charge density to the electrostatic potential $V(r)$ of the nucleus by
Poisson's equation,

$$\nabla^2 V(r) = 4\pi n \approx p_0^3$$

Our aim is to find the dependence of the radius R of the region that contains
the majority of electrons on the charge Z of the heavy atom. We suppose

$V(r)$ does not change in the region where there are a maximum number of electrons and replace the term $\nabla^2 V(r)$ by $\nabla^2 V(r) \approx \frac{V(r)}{R^2}$, where R is a measure of the region over which the charge density changes. The equation is dimensionally correct as the ∇ operator has the dimensions of $\frac{1}{R^2}$. The problem is how should R be chosen? Two conditions need to be satisfied,

(1) R should be chosen so that Poisson's equation, in this crude form, is satisfied.
(2) Also R should give a good estimate for the total charge Z of the atom since we would like

$$Z \approx nR^3 \approx (\nabla^2 V)R^3 \approx V(r)R$$

There are unknowns, $R, p_0, V(r)$ and two conditions. We need a third condition. The third condition is the energy. The energy at a given point of the atom when $p = p_0$. This energy E_0 using the free particle picture, is

$$E_0 = \frac{p_0^2}{2m} - V(r_0)$$

Next we make a physical observation that this value of E_0 is a constant throughout the atom. Again this makes sense in the free particle picture and also because if the energy changed there would be energy flows to adjust to the constant energy configuration. Finally we suppose that we can replace $V(r_0)$ by $V(r)$. Using these physical inputs and measuring $V(r)$ from $-E_0$, Poisson's equation gives

$$\nabla V(r) \approx V(r)^{\frac{3}{2}} \approx \frac{V(r)}{R^2}$$

From this it follows that $R \approx V^{-\frac{1}{4}}$ and thus we get $Z \approx V(r)R$ gives $R \approx Z^{-\frac{1}{4}}$. The length R thus describes the scale over which $V(r)$ changes. The result does not fix the length scale but gives its dependence on Z. The scale represents the transition region from where the majority of charges are located where $V(r)$ is expected to remain constant.

The method described is based on a number of physical ideas which we discussed. But the most important message of the example is that densely packed fermions can be treated as free particles. The approach described is called the Thomas-Fermi model.

2.1.5 *Estimating Lifetimes of Excited States*

We estimate the lifetime for a $2p \to 1s$ dipole transitions of a hydrogen atom. First we use dimensional analysis to find an atomic time scale. The time scale t should involve the energy difference of a transition ΔE and it should involve $\hbar$ as the process is quantum. Matching dimensions gives $t \approx \frac{\hbar}{\Delta E}$. For a $\Delta E = 10$ eV transition we get $t \approx 10^{-16}$ seconds. Now we use more physics to estimate the dipole lifetime of the 2p state of hydrogen. The inverse lifetime for the transition is given "Fermi's Golden Rule", which is,

$$\Gamma_{ij} = \frac{2\pi}{\hbar} \int |M_{ij}|^2 \frac{V d^3 k}{(2\pi)^3} \delta(E_i - E_j - \hbar\omega)$$

Let us check dimensions. We have the list

$$\hbar \to [E][T]$$
$$|M_{ij}|^2 \to [E]^2$$
$$\delta \to [E]^{-1}$$
$$\frac{V d^3 k}{(2\pi)^3} \to \text{dimensionless}$$

where $[E]$ is the dimension of energy. The δ function has the dimensions of the inverse of energy since $\int dE \delta(E) = 1$. Thus the dimension of Γ_{ij} is $[T^{-1}]$. We will use the atomic time scale as the unit of time. In the formula M_{ij} is the matrix element between states of energy E_i, E_j of the operator $M = -\frac{e}{mc}\mathbf{p}.\mathbf{A}$. with $\mathbf{p}, \mathbf{A}$ the momentum operator and the vector potential and $\hbar\omega$ is the energy of the emitted photon. We want to estimate Γ without doing integrals. We consider the one photon transition with $i = 1, j = 0$. Our estimates are,

$$M_{10} \approx \frac{e}{mc} \sqrt{\frac{2\pi\hbar c}{V\omega}} \, p_{10}$$
$$p_{10} \approx im\omega \, x_{10}$$
$$d^3 k = k^2 \frac{dk}{d\omega} d\Omega d\omega$$

where $d\Omega = \sin\theta d\theta d\phi$ the solid angle in polar coordinates (r, θ, ϕ). We explain how these estimates are made. The physics input is contained in the matrix element

$$M_{10} = [\mathbf{p}.\mathbf{A}]_{10}$$
$$\approx \mathbf{p}_{10}\mathbf{A}_{00}$$

where a complete set of state were used i.e. $(p.A)_{10} = \sum_n (p)_{1n}(A)_{n0} \approx (p)_{10}(A)_{00}$.

Let us explain how these two matrix elements can be estimated. First we estimate $x_{01} \approx \frac{\hbar^2}{Zme^2} = \frac{1}{Z}a_B$, which gives $p_{01} \approx \frac{ima_B\omega}{Z}$. This just says the matrix element is related to the orbit size and the transition frequency.

We next estimate A_{00}. To do this we use an unusual idea, namely that the matrix element comes from the fluctuating zero point electromagnetic field, required by quantum theory. This field has its associated vector potential. This is the idea we now follow.

The zero point field of momentum k has, by definition, energy density $E_{00} = \frac{\hbar\omega_k}{2V}$. It can be written, in terms of the electromagnetic field $(E_k)_{00}$, using wavenumber and frequency variables, and in turn the electromagnetic field itself can be written in terms of vector potentials. Thus we have,

$$E_{00} = \frac{(E_k)_{00}^2}{8\pi}$$

$$\mathbf{E_k} = \frac{-i\omega}{c}\mathbf{A_k}$$

where we have used the expression for the electromagnetic field in terms of the vector potential using frequency and wavenumber variables in the second equation. From this we get

$$E_{00} = \frac{\omega_k^2}{8\pi c^2}(|A_k|^2)_{00})]$$

$$= \frac{\hbar\omega_k}{2V}$$

Finally setting $\omega_k = ck$, valid for photons, we have our matrix element:

$$|(A_k)_{00}| = \sqrt{\frac{4\pi\hbar c^2}{V\omega_k}}$$

The fluctuating electromagnetic field is time dependent.

From the calculations we see that transitions from an excited state to a lower energy state, for an isolated atom, happen because of the presence of fluctuation zero point photons which have energy energy density equal to $\frac{1}{2V}\hbar\omega_k$. A truly remarkable result.

Putting these ideas together and using the following details:

$$kc = \omega, \ \delta(E_1 - E_0 - \hbar\omega)$$

$$= \frac{1}{\hbar}\delta(\frac{(E_1 - E_0)}{\hbar} - \omega)$$

and $a_B\omega \approx \frac{e^2}{\hbar}$ we get the expression of the lifetime for a $2p \to 1s$ transition for hydrogen atom as,

$$\Gamma_{01} \approx (\frac{e^2}{\hbar c})^3\omega = (\alpha)^3\omega$$

where $\alpha \approx 10^{-2}$, is the fine structure constant, which finally gives

$$t \approx \frac{1}{(\alpha)^3} t_a$$

where $t_a = \frac{1}{\omega} \approx 10^{-16}$, (an atomic time scale), so that the $2p \rightarrow 1s$ transition time $t \approx 10^{-10}$ sec for hydrogen. It is very interesting that excited states of hydrogen decay because of the presence of quantum virtual photon oscillations required by quantum theory. The matrix element $(|A_k|^2)_{00}$ used in the calculation contains no photons.

We give a few more examples of observable consequences of zero point photons. Three examples will be considered. The first estimates the Lamb shift, the second explains the Casimir force and the third predicts that interactions with zero point photons can create coherent nanoscale structures under suitable conditions.

2.2   The Lamb Shift

Let us look at the fine detailed structure of the hydrogen spectrum revealed by a quantum theory of radiation. A small shift in the $2S$ state of hydrogen relative to the $2P$ state was predicted and subsequently measured by Lamb. This is the Lamb shift. We will sketch a calculation of the Lamb shift due to Welton which is intuitive and uses the basic ideas of zero point energy of quantum theory.

The story behind Lamb's experiment is interesting. After Quantum Electrodynamics (QED) was constructed Uehling, in Columbia University, calculated the effect fluctuating radiation fields would have on the spectral lines of hydrogen. In QED the vacuum state, i.e. the state with no matter was seething with positive and negative charges. If a positive charge, such as a proton, was introduced in this vacuum it would be surrounded by a layer of negative charges. The vacuum should act as a dielectric. Thus Uehling felt that the 2S and 2P states of hydrogen should no longer have the same energy due to this shielding effect. The 2S state electron of hydrogen, which samples the centre of the charge, would thus experience a stronger attractive force and hence have stronger binding energy compared to the 2P state whose wave function vanishes at the charge centre. Uehling showed that this shielding effects modified the Coulomb in the following way:

$$V(r) = -\frac{e^2}{r} [1 + \frac{e^2 e^{-2mr}}{4\sqrt{\pi}(mr)^{3/2}}]$$

in units where $h = c = 1$. This modification predicted that 2S state would get lowered by 28 Megacyles/sec relative to the 2P state. Lamb set up to measure this tiny effect. When the experiment was done he found that indeed the 2S, 2P states had different energies but instead of the 2S having lower energy it had 1090 Megacycles/sec greater energy than the 2P state. This was the Lamb shift. A meeting in Shelter Island in 1947 resulted in an explanation of this result due to Bethe [19]. The hotel where the meeting took place has a plaque with a list of those who attended the meeting. It was a historic event and marked a great triumph of theoretical and experimental physics. Let us now sketch Welton's calculation.

The idea is that as the electron orbits the proton its location fluctuates from its average classical value due to fluctuating vacuum quantum radiation field. This change in position changes the Coulomb potential and hence the energy of the electron. The fluctuation in position come from the fluctuating radiation field. As the radiation field can be regarded as a collection of oscillators these vacuum fluctuations have energy given by the zero point energy of each allowed frequency mode of the photons. Let us give some details. The position fluctuation of an electron $\delta_i(t)$ satisfies the equation

$$m \frac{d^2 \delta_i(t)}{dt^2} = eE_i(t)$$

where $E_i(t)$ is the fluctuating radiation electric field due to zero point energy,(we have suppressed $\vec{x}$ dependence of both δ and E). m the electron mass, e the electron charge. Taking Fourier transforms this becomes (again these Fourier transforms are for $t, \vec{x}$)

$$-m\omega^2 \delta_i(\omega) = eE_i(\omega)$$

Squaring this gives

$$m^2 \omega^2 \delta^2(\omega) = e^2 E^2(\omega)$$

The presence of this fluctuation modifies the Coulomb potential $V(r)$ at any r value to $V(r + \delta(t))$. Taylor expanding we get

$$V(r + \delta(t)) = V(r) + \delta_i(t)\partial_i V(r) + \frac{1}{2}\delta_i(t)\delta_j(t)\partial_i\partial_j V(r) + \dots$$

where the summation convention has been used. If we assume $\delta_i(t)$ is isotropic then the time average of this expression becomes

$$< V(r + \delta) - V(r) > = \frac{1}{6} < \delta^2(t) > \nabla^2 V(r) + \dots$$

where we set $< \delta_i(t)\delta_j(t) >= \frac{1}{3}\delta_{i,j} < \delta^2(t) >$, where $\delta_{i,j}$ is the Kronecker delta function. At this stage we can think of $< V(r+\delta) - V(r) >= \Delta V$ as a perturbation. This changes a hydrogen level energy E_n by an amount ΔE_n given by

$$\Delta E_n = < \psi_n | \Delta V | \psi_n >$$

where ψ_n is the wave function of the electron in the energy state $E - n$ and $< \psi_n | \Delta V | \psi_n >= \int d^3 r \psi_n(r)^\dagger \Delta V \psi_n(r)$. At this stage we use four facts

(1) The space-time average of $\delta^2(t)$ is equal to its momentum- frequency average of $\delta^2(\omega)$. (We have suppressed the space dependence of $\delta(t)$ and the momentum dependence of $\delta(\omega)$).
(2) $\nabla^2 V(r) = 4\pi e^2 \delta^3(r)$, where $\delta^3(r)$ is the three dimensional Dirac delta function.
(3) The density of states $\rho(\omega)$ for free photons of frequency ω is $\rho(\omega) = \frac{V_0 \omega^2}{2\pi^2 (ch)^3}$, where c is the velocity of light and V_0 is the volume.
(4) The square of the fluctuating frequency representation of the electric field $E(\omega)$ is related to the zero point energy of the radiation field by the formula $\frac{E(\omega,\vec{k}))^2}{8\pi} = \frac{1}{2}\frac{\hbar\omega\delta}{V_o}(\hbar\omega - c|\vec{k}|)$ where $\delta(\hbar\omega - c|\vec{k}|)$ is the Dirac delta function introduced to implement the energy momentum relation for real photons.

Putting these details in we get our expression for ΔE_n as an integral over ω, namely

$$\Delta E_S = \int d\omega \delta(\omega)^2 (\nabla^2 V(r)) |\psi_S(r)|^2$$

$$= \frac{e^4}{3h^3 c^3 \pi} |\psi_S(0)|^2 \int \frac{d\omega}{\omega}$$

where $|\psi_S(0)|^2$ is the square of the absolute value of the S state wave function at $r = 0$. There is no shift for the P state energy as its wave function is zero at $r = 0$. The lower limit of the integral is taken to be the frequency of the Bohr atom for $n = 1$ while the upper limit is taken to be $\frac{mc^2}{h}$. From the point of mathematical physics the expression for the density of states used contains in it is a deep mathematical result. Namely it tells us that the density of eigenstates of the Laplacian operator, present in the hydrogen atom problem, is related to the geometry of the enclosure being used. The study of the precise way in which the asymptotic form of the density of eigenstates of the Laplacian operator is fixed by geometry of the domain in which the eigenvalue problem is being solved is the subject matter of spectral geometry. The deep result from this field is that the leading term for the density of eigenstates of the Laplacian operator is proportional to the volume of the region in which we are interested. The next term depends on the surface area. The coefficients of proportionality are universal. Thus if we were interested, for instance, in working out the Lamb shift in a small cavity we should include the surface term contribution to properly represent the density of states.

We next turn to a totally unexpected result predicted by taking the zero point photon field seriously: the Casimir Effect.

2.2.1   *The Casimir Effect*

In 1948 Casimir made an astonishing discovery. He showed that an attractive force between two conducting parallel plates was predicted to exist from quantum theory. The origin of the force was due to the zero point energy of the vacuum electromagnetic field. This force, of quantum origin, has now been measured with great accuracy. It is observed. A quick way of understanding the force is to consider a one dimensional system of length L. If the free electromagnetic field to taken to be waves oscillating in this cavity that vanish at the end points then the classical oscillation frequencies of the system are: $\omega_n = \frac{nc\pi}{L}$. Each classical frequency represents a harmonic oscillator so that the associated zero point energy of the system is

$$E = \sum_n \hbar \frac{nc\pi}{2L}$$

If the "volume" L is changed the energy changes and there is a force. The sum is, of course, divergent but there are many ways of handling this difficulty. A fast way is to think of the sum $\sum n$ as $\sum_n \frac{1}{n^s}$ which is the famous zeta function $\zeta(s)$ of Riemann. The zeta function is an analytic function of s which can be continued to the point $s = -1$ where $\zeta(-1) = -\frac{1}{6}$. This procedure thus gives $\sum n = -\frac{1}{6}$ so that $E = -\frac{c\pi}{12L}$ from which the attractive force $F = -\frac{dE}{dL} = -\frac{c\pi}{12l^2}$ is obtained. A generalisation of this procedure can be used to calculate the attractive force between two conducting parallel plates.

2.2.2   *Zero Point Effects for Nanoscale Structures*

The result that zero point EM fields change the positions of orbiting electrons was established when we discussed the Lamb shift. Such a positional change implies an associated volume change for an assembly of such orbits. Such an induced volume change, we show, generates an induced EM force. The presence of such a force can, we show, lead to the formation of coherent nanoscale structures. This is a speculative theoretical idea which has to be checked by further experiments. To establish the result also requires a number of additional theoretical assumptions, which we discuss later. Thus the class of nanosystems for which coherent structure formation is expected is limited. However for nano systems belonging to this class the spontaneous formation of coherent nanoscale structures is predicted. Such structures are expected to have unusual physical properties.

We establish this result for surface nanobubbles on hydrophobic surfaces that have been observed but the results described are expected to hold for more general systems.

Let us start by first summarise some experimental results about surface nanobubbles in water. Stable gas surface nanobubble have been observed to form on hydrophobic surfaces in water. They are remarkably stable even when the water temperature is increased to close to the boiling point. The nanobubbles have lifetimes of the order of 10^9 sec or more and have diameters of $\approx$ 100nm. Understanding the stability and long lifetime of surface nanobubbles is a challenge as the Laplace pressure inside such a small object should, by diffusion, force out all the gases in them leading to their collapse in microseconds.

We show that the zero point EM ideas sketched can provide an explanation of what is observed.

Consider surface nanobubbles in water with a well defined volume V_c. This volume contains a collection of electrons in orbital motion. We examine the effect of zero point EM fields on such an assembly.

Let us analyse the system. The surface layer of the nanobubble contains a collection of N electrons in electronic orbits. A well defined surface layer volume thus can be defined. The vacuum electromagnetic field couples to all these orbiting electrons, each of which has mass m and charge e. We have already worked out the shift in position $< \delta^2 >$ for a single orbiting electron due to zero point EM fields. Let us make use of that calculation for our assembly. Let $\vec{E}(t,\vec{x})$ denote, as before, the time dependent electromagnetic field due to the vacuum fluctuation and $\vec{\delta_i}$ denote the fluctuation in the position of the ith electron due to the effect of $\vec{E}(t,x)$. Then we have

$$m \sum_i \ddot{\vec{\delta_i}}(t,\vec{x}) = Ne\vec{E}(t,\vec{x}), \quad i = 1,2,.....N.$$

Taking the Fourier transform, we get

$$-m \sum_i \omega^2 \vec{\delta}_i(\omega, \vec{x}) = Ne\vec{E}(\omega, \vec{x}),$$

where we have used the same symbols $\vec{\delta}$ and $\vec{E}$ for the Fourier transforms of these quantities. Thus we get

$$\left| \sum_i \vec{\delta}_i(\omega, \vec{x}) \right|^2 = N^2 \frac{e^2 E^2(\omega, \vec{x})}{m^2 w^4},$$

where $E^2 \equiv |\vec{E}|^2$.

We now calculate the time average of the fluctuations. We assume that the fluctuations are independent so that the cross terms vanish and that the time averaged fluctuation $< |\vec{\delta}_i|^2 > \equiv < \delta^2 >$ are the same for all the electrons. From these assumptions and using the result for the value of δ^2 found earlier we get our final expression for the average squared fluctuation of the position of an electron when it is part of an assembly of N electrons as

$$< \delta^2 >_N = \frac{2N\alpha\lambda_c^2}{\pi} \ln \frac{1}{\alpha^2} \equiv \delta_N^2,$$

where $\alpha = \frac{e^2}{\hbar c}$ is the fine structure constant and $\lambda_c = \frac{\hbar}{mc}$ is the Compton wavelength of the electron. The factors in the log term come from limits of ω integration which are taken to be $\frac{mc^2}{\hbar}$ and $\frac{me^4}{\hbar^3}$ representing a relativistic cut off set by the electron mass and a lower frequency cut off set by the atomic scale Bohr frequency. We will set $\ln \frac{1}{\alpha^2} \approx 8$. Thus $< \delta^2 >_N \approx \frac{16N\alpha\lambda_c^2}{\pi}$. This is a constant universal expression. The important point to note is that the average positional shift per electron is modified by a factor of $\sqrt{N}$ when the electron belongs to the assembly.

We next show how this result can be used to determine the coupling of a charge to the zero point field when it is part of an assembly of charges close together. The idea is that the fluctuation-induced position change δ_N produces a change of the well defined nanovolume $V_c \approx l^3$ and this changes the zero point energy of the volume and leads to a force. We saw how such an idea led to the observed Casimir force. The volume of the nanoshell we assume is set by the wavelength λ corresponding to a transition of energy ≈ 10 eV. We set $l = \lambda$ relating the transition wavelength to the coherence length. This induced force tells how the vacuum EM field interacts with a charge belonging to the assembly.

To determine this force. we start with our well-defined nanocluster of orbiting charged particles in volume V_c. The energy density of the fluctuating field in this volume is $|\vec{E}|^2 = \frac{1}{2V_c} \hbar\omega$ for frequency ω, where $\vec{E}$ is the vacuum EM field, given by $\vec{E} = \vec{u} \frac{\sqrt{\hbar\omega}}{\sqrt{2V_c}} e^{-i\omega t}$. Consider the effect of volume change due to the position change $\sqrt{<\delta^2(t,\vec{x})>_N} \approx 4\sqrt{\frac{\alpha\lambda_c^2}{\pi}N}$ on $\vec{E}$ The volume change produces, as we now show, an induced EM force, which we write as $e\vec{E}_f(t)$.

We determine $\vec{E}_f(t)$ in two steps. In the first step we define $\vec{E}_f$ by the equation

$$\sqrt{\alpha}\vec{E}_f(t) = \vec{u}\sqrt{\hbar\omega}[\frac{1}{\sqrt{2(V_c + \delta V)}} - \frac{1}{\sqrt{2V_c}}]e^{-i\omega t}$$

where $\frac{\delta V_c}{V_c} = \frac{3\sqrt{<\delta>^2}}{l}$. A constraint to remember is that we must have $\frac{\delta V}{V} << 1$. Thus we get

$$e\vec{E}_f(t) = \vec{u}\,\frac{3}{2}\sqrt{\frac{16c}{\pi}\frac{N\alpha\lambda_c^2}{2V_c\omega}}(\frac{1}{l})\hbar\omega\ e^{-i\omega t}.$$

The fluctuating EM force $e\vec{E}_f$ can now be used to produce an interaction energy term, $e\vec{E}_f(t).\vec{x}$ with an electron, located at $\vec{x}$ which is simply the usual $\frac{e}{c}\vec{j}.\vec{A}$ term, where the vector potential $\vec{A}$ is defined by $\vec{E}_f = \frac{1}{c}\frac{\partial\vec{A}}{\partial t}$. Thus the induced EM term constructed is a standard field-current interaction and can cause a transition between states $|i>, |f>$. Its transition matrix element is given by the expression

$$< i|e\vec{E}_f.\vec{x}|f > = \frac{< i|(\vec{x}.\vec{u})|f >}{\sqrt{2V_c\omega/c}}\frac{3}{2}\sqrt{\alpha}\sqrt{\frac{16N\lambda_c^2}{\pi l^2}}\ \hbar\omega$$

and we shall use $r =< i \mid \vec{x}, \vec{u} \mid f >$ henceforth. The expression for the vacuum field induced transition amplitude represents a coupling between two electronic states and the zero point photon induced em field. It has the structure of the usual dipole transition but with a universal multiplicative factor $F = \sqrt{\alpha\frac{16N\lambda_c^2}{\pi l^2}}$ due to the fact that the interaction is induced from zero point fluctuations. The condition $\frac{\delta V_c}{V_c} < 1$ tells us that $F < 1$. Thus to make the transition large we should choose F to be as large as possible but satisfy the condition $F < 1$. Setting N is $\approx 10^7$ is such a choice. For such a choice the mixing of states takes place with high probability. This is what can happen in the nanobubble layer..

Another important result that can be extracted from the transition matrix element is a frequency relation between a collective frequency Ω and the transition frequency ω, which is,

$$\Omega = G\omega,$$

$$G = \sqrt{\alpha}\ r\ \frac{3}{2}\ \sqrt{\frac{Nc}{2V_c\omega}}\ \left(\sqrt{\frac{16\lambda_c^2}{\pi l^2}}\right).$$

We note that $< i|e\vec{E}_F.\vec{x}|f >$ is a characteristic energy associated with the transition which we have written as $\hbar\Omega$. Finally we show that the mixing of states described can lower the ground state energy. To do this we focus on just two states of our system that get mixed by the induced zero point EM field. The ground state and one excited state. The induced electric field is independent of $\vec{x}$ and is small. Hence a simple perturbation treatment of its effect is allowed. We describe energies in terms of frequencies $\Omega, \omega_1, \omega_2$, of our two states. Consider the Hamiltonian

$$\frac{H}{\hbar} = \begin{pmatrix} \omega_1 & \Omega \\ \Omega & \omega_2 \end{pmatrix}$$

which acts on the states characterized by eigenvalues ω_1 and ω_2 and the interaction term mixes the two states. The eigenvalues of the interaction Hamiltonian H are given by

$$\lambda_\pm = \frac{(\omega_1 + \omega_2)}{2} \left[1 \pm \sqrt{1 + \frac{4(\Omega^2 - \omega_1\omega_2)}{(\omega_1 + \omega_2)^2}} \right].$$

For $\Omega^2 > \omega_1\omega_2$ we see that one of the eigenvalues is

$$\lambda_- \approx -\frac{\Omega^2 - \omega_1\omega_2}{(\omega_1 + \omega_2)} < 0.$$

Setting the ground state $\omega_1 = 0$, we see that the binding energy per water molecule is thus $\approx G^2\omega$. For nanobubbles in water we must take into account one important fact. The excited states of water are known to be unstable in bulk water with lifetimes in the femtosecond range. Thus stable coherent structures cannot form. However there is also evidence that water layers, of a few molecular size thickness, next to a surfaces can have stable excited states. Thus for nanobubbles we set $V_c = b\lambda^2$ where $b \approx 10^{-7}$ cm. We have taken the layer thickness to be three molecules. A consequence of this step is that it modifies the expression for G. Now

$$G = \frac{\sqrt{\alpha}\, r}{\sqrt{\pi}} \sqrt{\frac{Nc}{2V_c\omega}} \frac{4\lambda_c}{b}$$

The reason for the change is because now

$$V_c + \delta V_c = (l + \delta)^2(b + \delta)$$

A value of G of $\approx 6 \times 10^{-2}$ is possible for suitably chosen values for $N \approx 10^7, r \approx 6 \times 10^{-7}$ cm, $\omega \approx 10^{16}s^{-1}$, $V_c \approx \times 10^{-17}$ cc, $b \approx 10^{-7}$. Recall we have assumed the coherent length scale $l = \lambda$, where λ is the wavelength associated with the frequency ω. The large value for r reflects the size of a weakly bound state. This estimate can be understood as follows: For a single electronic excitation of binding energy the asymptotic large r form for of the radial wavefunction $R(r)$ is $R(r) \approx e^{-r\alpha}r^\beta$ where $\alpha = \frac{\sqrt{2mE}}{\hbar}$. The value of r for which the radial wavefunction reaches its maximum vale is $r \approx \frac{\beta}{\alpha}$. We take this value of r to represent the excitation orbit size. The parameter β is expected to be an integer greater than one for an excited state, m is the mass of the electron. For a binding energy of 0.1 eV and $\beta = 5$ gives the estimate written down. The lowering of the ground state energy gives a binding energy per molecule $b = G^2\omega$. Thus the binding is greatest for large ω which corresponds to a transition from the ground state to an energy state close to ionisation i.e. to an electron state that is weakly bound. This result predicts that the coherence property of the nanoscale structure should remain as long as the thermal energy $kT < G^2\omega$, where T is the temperature and k the Boltzmann constant. For a value of $G^2\omega \approx .05$ eV means that nanobubbles should remain stable, even when the water is raised to temperatures close to boiling temperature. This surprising result that nanobubble remain stable at high temperatures has been observed.

Let us now highlight the simplifying assumptions made. We assumed that the electrons even when in a cluster can be described by stable bound states with energy eigenvalues that are known and that a well defined starting volume can be introduced for the system. For a nanoscale solid such a system a volume can be easily be identified. For nanobubbles in water we took the observed surface nanobubble sizes as an indication of the size that is relevant. These sizes are typically 100nm. Such a size corresponds to a transition wavelength of $\lambda \approx 12$ eV. The ionisation energy of water is 12.6 eV. Thus the numbers gathered from experiment suggest that we take the coherent volume length scale l to be proportional to λ. We have set $l = \lambda$ in order to get sizes of nanobubbles that are observed. However for water we know that its excited states are unstable but that surface interactions can stabilise the excited states for a few layers of molecules close to the surface. For this reason we wrote $V_c = b\lambda^2$ with $b \approx 10^{-7}$ a three molecular layer of water for which the excited states are stabilised.

Although we have discussed the coherent structure formation for surface nanobubble it should be clear that the method is applicable to a large class of nanoscale structures for which the assumptions listed hold.

We next explain how coherence helps us understand the long lifetimes of surface nanobubbles. The reason is coherent structures reduce the value of the diffusion coefficient D by a factor of 10^10. This result is easily understood since the process of gas diffusion depends on the diffusion coefficient D which is inversely proportional to the total scattering cross section σ_T of one trapped gas with the N water molecules present. Now $D = <v> L$, where $<v>$ is the average thermal speed of the gas molecules and L, the mean free path for gas molecules interacting with water molecules, is given by $L = \frac{1}{\sigma_T}$ where there are N water molecules present in the volume and σ_T is the total scattering cross section of the gas molecule with the N water molecules present. For incoherent scattering $\sigma_T \approx N\sigma$, while for coherent scattering $\sigma_T \approx N^2\sigma$ where σ is the cross-section for a single gas molecule-single water molecule scattering. Thus if the nanosurface layer is made up of the coherent molecules the value of D is lowered by a factor of $N \approx 10^9$ and their lifetime is greatly enhanced leading to a much longer lifetime.

We used an estimate for gas diffusion through water to explain the long life time for nanobubbles. Now we justify the result used using qualitative arguments.

2.2.3 *Estimating the Diffusion Coefficient*

The diffusion coefficient D was estimated by the expression $D \approx l <v>$, where l was the free free path for a gas molecule through water and $<v>$ its thermal velocity. Let us understand this approximation by following a series of steps. In the first step we use dimensional analysis for this we need to determine the dimension of the diffusion coefficient and then list variables on which it should depend. To determine the dimension of the diffusion coefficient D we turn to the diffusion equation, in the following way,

$$\frac{\partial y(x,t)}{\partial t} = D\nabla^2 y(x,t)$$

The dimensions of the operator on the left hand side is $\frac{1}{[T]}$ while the operator dimension of the left hand side is $\frac{[D]}{[L^2]}$, for the dimensions to match the dimension of the diffusion coefficient $D = [D] = \frac{[L^2]}{[T]} = \frac{[L]}{[T]}[L] = [v][L]$ where $[v]$ is the velocity dimension, $[L]$ the length dimension and $[T]$ the time dimension. Thus from dimensional analysis we see that the diffusion coefficient should be related to a velocity multiplied by a length. The appropriate velocity and length scales then to be determined using a model. In the simplest model the diffusion coefficient is taken to dependent on the way the motion of gas molecules is slowed down by collisions with water molecules and the speed with which the gas molecule is moving. A crude measure of the effect of collisions is given by the mean free and a crude measure of its motion is given by the thermal speed of the gas molecules. The mean free path can be estimated from a knowledge of molecules present in the volume and the scattering cross section of the gas molecules with water molecules. Thus we want to construct a length scale from $\frac{N}{V}$ and σ where σ the cross section has the dimensions of $[L]$ and $\frac{N}{V} = N_0$ has dimension $[L^3]$. From this we conclude that the mean free path $l = \frac{1}{N_0 \sigma}$. The simple argument, suitably generalised, can be used to estimate many transport coefficients.

If we wish to proceed to a level of greater exactness we need to use the Boltzmann transport equations to determine D.

2.2.4   *Kolmogorov's Law for Turbulence*

We now show how qualitative reasoning can be used to understand Kolmogorov's scaling law for turbulence. Turbulent flow occurs in fluids when the velocity changes between fluid layers changes quickly over a short distance scale. In turbulent flows the fluid particles seem to move with random speeds with different wavelengths of eddies forming. These features of turbulence were discovered by Reynolds who also discovered a numerical factor, now called the Reynolds number, which could predict when turbulent flow is expected. The Reynold's number R is defined as $R = \frac{L \Delta V}{\eta}$, where L is a characteristic length scale over which the magnitude of the velocity change is ΔV and η is the kinematic viscosity. Turbulence occurs when R is large.

Kolmogorov in 1941 made an astonishing discovery that hidden in the seemingly random flow there was hidden a simple scaling law that governed the way energy of the fluid waves is distributed over different wavelengths. Instead of using wavelengths Kolmogorov's law is usually stated in terms of wave numbers $k = \frac{1}{\lambda}$ where λ represents wavelength. Kolmogorov's law states that the energy $E(k)$ of the fluid with wavenumber k obeys the simple scaling law $E(k) \approx k^{\frac{5}{3}}$.

Let us understand this result by using the qualitative reasoning of Kolmogorov. The starting point of the approach is an idea due to Fry Richardson who had suggested that turbulence was a way of transferring energy from a large scale through the break up of large scale circulations to smaller and smaller eddies. Subsequently it was established that although turbulence resulted in the dissipation of energy dissipation due to viscosity only occur at the very short scale. There was thus a range of wave numbers for which there was no energy dissipation but simply energy transfer from a large scale to a small scale. This range of wave numbers is called the inertial range. Kolmogorov's law is expected to hold in this range. Let us list the physically relevant variables of the fluid particles that are in the inertial range at step n of the process of energy transference. We have

$$\text{energy } E_n \approx v_n^2$$
$$\text{time } \tau_n \approx l_n v_n$$
$$\text{energy transfer } \epsilon_n \approx \frac{E_n}{\tau_n}$$

The assumption made by Kolmogorov was that ϵ_n was scale invariant i.e. it did not depend on the eddy scale label n. Each of the expressions is the simplest dimensionally correct possibility. Using Kolmogorov's assumption we get

$$v_n^2 \approx \epsilon.\tau_n$$
$$\approx \epsilon \frac{l_n}{v_n}$$
$$v_n^3 \approx \epsilon l_n$$

From this we get $v_n^2 \approx \epsilon l_n^{\frac{2}{3}}$. Taking Fourier transforms we get

$$E(k) \approx \epsilon \int dy e^{iky} y^{\frac{2}{3}}$$

changing variables to $z = ky$ we get

$$E(k) \approx k^{-\frac{5}{3}}$$

which is Kolmogorov's result.

2.3 Turbulence in Graphene

We now examine a rather unexpected theoretical result namely that current flow in graphene could be turbulent. This example is included because of its novelty and also because the method used to derive the result could be used in situations where scaling behavior is expected. The possibility of turbulent current flow is implied as soon as it was shown that the ratio of the shear viscosity η to the entropy density s of current flow was shown to be small. This result was established in the hydrodynamic region, where the current flow is described by a quantum Boltzmann equation and that turbulent flow might occur as a hydrodynamic system with low viscosity is

expected to be turbulent. A turbulent regime is characterized by a large
value for the Reynold's number, which for graphene at a given temperature
T was shown to be,

$$Re = \frac{s/k_B}{\eta/\hbar} \frac{k_B T}{\hbar v/L} \frac{u_{typ}}{v},$$

where v is the Fermi velocity, u_{typ} denotes a typical velocity and L is the
characteristic length scale for the velocity gradients. The kind of turbulence
expected is novel. It is quantum turbulence.

A quantum turbulent state is possible for a complex quantum system
with many degrees of freedom. It represents a highly excited state of the
quantum system where energy is pumped in one scale and dissipated due to
nonlinearities present at a different scale. Turbulent flows are characterized
by a probability distribution function (PDF) with scaling properties and an
associated non-vanishing energy flux. Quantum turbulent flows have been
studied using the methods of quantum field theory, as such systems have
many degrees of freedom exhibiting stochastic behavior. Systems where the
wave motion is a recognized feature and turbulent flows are generated due
to nonlinear interactions are known as weak wave turbulent systems.

Our starting point will be to describe graphene by a nonlinear quantum
Hamiltonian and follow the well established methods of weak wave turbulent flows to determine the turbulent PDF for graphene and calculate the
contribution made by this component to conductance.

Let us sketch how we can calculate a possible turbulent flow probability distribution function (TPDF) for graphene in the hydrodynamic regime
treating it as a weak wave system. The wave motion comes from the free
Dirac equation. The nonlinearity in graphene is due to the electron-electron
Coulomb interaction, which is described by a term quartic in the field operators. We show that these combined types of interactions admit a solution
with a turbulent flow. Using the TPDF we then calculate the conductance
due to this flow. We find that the contribution to the conductance due to
turbulent flow has a structure different from that found using scattering.

Graphene near a Fermi point is described by the Hamiltonian,

$$H = H_0 + H_1,$$

$$H_0 = \sum_{\lambda,i} \int \frac{d^2k}{(2\pi)^2} \lambda v_F k a^\dagger_{\lambda i}(\mathbf{k}) a_{\lambda i}(\mathbf{k}),$$

$$H_1 = \sum_{\lambda_1,\lambda_2,\lambda_3,\lambda_4;i,j} \int \frac{d^2k_1}{(2\pi)^2} \frac{d^2k_2}{(2\pi)^2} \frac{d^2k_3}{(2\pi)^2} \frac{d^2k_4}{(2\pi)^2} \delta^2(\mathbf{k_1} + \mathbf{k_2} - \mathbf{k_3} - \mathbf{k_4})$$

$$\times T_{\lambda_1\lambda_2\lambda_3\lambda_4}(\mathbf{k_1}, \mathbf{k_2}, \mathbf{k_3}, \mathbf{k_4}) a^\dagger_{\lambda_4 j}(\mathbf{k_4}) a^\dagger_{\lambda_3 i}(\mathbf{k_3}) a_{\lambda_2 i}(\mathbf{k_2}) a_{\lambda_1 j}(\mathbf{k_1}),$$

where a_{+i}, a_{-i} are the Fourier mode operators, $i, j = 1, ..., 4$ for graphene corresponding to two valleys and two spins and $\lambda = (+, -)$. We also define $k = |\mathbf{k}|$, which is the modulus of the two momentum $\mathbf{k}$. The explicit expression for $T_{\lambda_1\lambda_2\lambda_3\lambda_4}(\mathbf{k_1}, \mathbf{k_2}, \mathbf{k_3}, \mathbf{k_4})$. For our discussion, we need to note that under scaling,

$$T(\mu\mathbf{k_1}, \mu\mathbf{k_2}, \mu\mathbf{k_3}, \mu\mathbf{k_4}) = \frac{1}{\mu} T(\mathbf{k_1}, \mathbf{k_2}, \mathbf{k_3}, \mathbf{k_4}),$$

where μ is a scalar.

A turbulent flow represents a steady state of the system in which, unlike a thermal equilibrium state, there exists a transport of a charge current. Such flows are expected to have scaling properties. The basic idea is to calculate

$$\frac{d}{dt} n_{k,\lambda} = \frac{d}{dt} < a^\dagger_{k,\lambda} a_{k,\lambda} >$$

between two "in-states", which are eigenstates of H_0. Next we look for scaling solutions which have the property that

$$\frac{d}{dt} n_{k,\lambda} = 0$$

and have a nonvanishing associated flux, to be discussed later. From Eq. (4) we find that

$$0 = \sum \int \frac{d^2k_1}{(2\pi)^2} \frac{d^2k_2}{(2\pi)^2} \frac{d^2k_3}{(2\pi)^2} \delta^2(\mathbf{k_1} + \mathbf{k_2} - \mathbf{k_3} - \mathbf{k}) |T_{\lambda_1\lambda_2\lambda_3\lambda_4}(\mathbf{k_1}, \mathbf{k_2}, \mathbf{k_3}, \mathbf{k})|^2 n_{k_1} n_{k_2} n_{k_3} n_k$$

$$\times \left[-16 \left(\frac{1}{n_k} + \frac{1}{n_{k_3}} - \frac{1}{n_{k_1}} - \frac{1}{n_{k_2}} \right) + 8 \left(\frac{1}{n_{k_3} n_k} + \frac{1}{n_{k_1} n_k} - \frac{1}{n_{k_2} n_{k_3}} - \frac{1}{n_{k_1} n_{k_2}} \right) \right].$$

We would like to obtain the stationary solutions in terms of the energy instead of the momenta. We shall further assume that the solutions are isotropic, so that both $n_{k,\lambda}$ and the energy ϵ depend only on the magnitude k of the momentum vector $\mathbf{k}$. Following the procedure of Zakharov we change variables from k to $\epsilon(k)$, where the latter denotes the energy and replace T by

$$U(\epsilon_1\epsilon_2\epsilon_3\epsilon) = (k_1k_2k_3k)\left|\frac{d\epsilon_1}{dk_1}\frac{d\epsilon_2}{dk_2}\frac{d\epsilon_3}{dk_3}\frac{d\epsilon}{dk}\right|^{-1}\int|T_{k_1k_2k_3k_4}|^2\delta^2(k_1+k_2-k_3-k)d\Omega_1 d\Omega_2 d\Omega_3,$$

where we have used the relation $d^2k_i = k_i dk_i d\Omega_i$. From Eq. (6) it can be easily seen that the quantity U is invariant under scaling.

From Eq. (5) and Eq. (6) and using the free particle density, we get

$$0 = \int_0^\infty d\epsilon_1 d\epsilon_2 d\epsilon_3 U(\epsilon_1\epsilon_2\epsilon_3\epsilon)\delta(\epsilon_1+\epsilon_2-\epsilon_3-\epsilon)n_{k_1}n_{k_2}n_{k_3}n_k$$

$$\times\left[-16\left(\frac{1}{n_k}+\frac{1}{n_{k_3}}-\frac{1}{n_{k_1}}-\frac{1}{n_{k_2}}\right)+8\left(\frac{1}{n_{k_3}n_k}+\frac{1}{n_{k_1}n_k}-\frac{1}{n_{k_2}n_{k_3}}-\frac{1}{n_{k_1}n_{k_2}}\right)\right].$$

We now analyze Eq. (7). The δ-function can be used to carry out the ϵ_3 integral. What remains is a region D of the (ϵ_1,ϵ_2) plane. In addition we have a constraint that $\epsilon_3 = \epsilon_1 + \epsilon_2 - \epsilon \geq 0$. We now divide D into four sectors as follows

$$D_1 = \{(\epsilon_1,\epsilon_2)\in D \mid \epsilon_1 < \epsilon,\ \epsilon_2 < \epsilon\}$$
$$D_2 = \{(\epsilon_1,\epsilon_2)\in D \mid \epsilon_1 > \epsilon,\ \epsilon_2 > \epsilon\}$$
$$D_3 = \{(\epsilon_1,\epsilon_2)\in D \mid \epsilon_1 < \epsilon,\ \epsilon_2 > \epsilon\}$$
$$D_4 = \{(\epsilon_1,\epsilon_2)\in D \mid \epsilon_1 > \epsilon,\ \epsilon_2 < \epsilon\}$$

It is possible to map the regions D_2, D_3, D_4 into D_1 using transformations called the Zakharov transform. For example D_2 with variables $(\epsilon_1',\epsilon_2')$ can be mapped to D_1 with variables (ϵ_1,ϵ_2) using

$$\epsilon_1' = \frac{\epsilon\epsilon_1}{\epsilon_1+\epsilon_2-\epsilon}$$
$$\epsilon_2' = \frac{\epsilon\epsilon_2}{\epsilon_1+\epsilon_2-\epsilon}$$

and so on. Using these transformations and the ansatz $n(\epsilon) = C\epsilon^{-x}$, we get

$$0 = \int_{D_1} d\epsilon_1 d\epsilon_2 U(\epsilon_1,\epsilon_2,\epsilon_1+\epsilon_2-\epsilon,\epsilon)(\epsilon_1\epsilon_2(\epsilon_1+\epsilon_2-\epsilon)\epsilon)^x$$

$$\left[1+(\frac{\epsilon_1+\epsilon_2-\epsilon}{\epsilon})^y-(\frac{\epsilon_2}{\epsilon})^y-(\frac{\epsilon_1}{\epsilon})^y\right]\times \text{other factors},$$

where $y = 3x - 3$. The factor within the square brackets in Eq.(10) can be made to vanish by choosing $y = 0$ or $y = 1$, irrespective of other factors in the equation. It is possible to show that only the solution corresponding to $y = 0$ or $x = 1$ has an associated nonzero flux and hence represents a turbulent state. This is the solution that we shall use.

Our solution for the TPDF is thus given by

$$n(\epsilon) = \frac{C}{\epsilon} = \frac{C}{\hbar k},$$

where for the moment we have set $v_F = 1$. We now need to determine C. To do this, we observe that the total number of turbulent particles per unit area, that is the turbulent number density N_T is given by

$$N_T = \frac{C}{2\pi} \frac{k_{max}}{\hbar},$$

We find

$$C\epsilon_{max} = N_T v_F^2 \left[\frac{(2\pi\hbar)^2}{4} \frac{1}{2\pi} \right]^{-1},$$

where $\epsilon_{max} = v_F k_{max} = \gamma k_B T$, where T is the background temperature of the system before the turbulent flow starts, k_B is the Boltzmann constant and $\gamma << 1$ is a constant, which is required for the hydrodynamic regime to be relevant.

We now calculate the conductance using the quantum Boltzmann equation. For this purpose we introduce a time dependent electric field $\mathbf{E}(t)$, which is assumed to be small. We write the time dependent density

$$f_\lambda = < a^\dagger_{\lambda,i}(k,t) a_{\lambda,i}(k,t) >$$

and $f_\lambda^0 = \frac{C}{k}$, the turbulent time independent PDF and determine f_λ by using the equation

$$\left(\frac{\partial}{\partial t} + \mathbf{E} \cdot \frac{\partial}{\partial \mathbf{k}} \right) f_\lambda = 0.$$

We use the ansatz

$$\tilde{f}_\lambda(k,\omega) = \tilde{f}_\lambda^0(k) + \frac{\mathbf{E} \cdot \mathbf{k}}{k} \tilde{g}_\lambda(k,\omega),$$

where

$$\int \tilde{f}_\lambda(k,\omega) e^{i\omega t} \frac{d\omega}{2\pi} = f_\lambda(k,t).$$

We find

$$\tilde{f}_\lambda = f_\lambda^0 + \frac{\mathbf{E} \cdot \mathbf{k}}{k} \left(\frac{\partial f_\lambda^0}{\partial k} \right) \frac{1}{(-i\omega + 0_+)},$$

where we have ignored the terms higher order in $\mathbf{E}$. It is now a simple matter to calculate the conductance. Taking the electric field to be in the x direction, we have

$$\sigma_T(\omega) = \frac{<J_1>}{E_x},$$

where $(<J_1>) = f_\lambda$ and σ_T is the contribution of the turbulent flow to the conductivity. Substituting for f_λ and introducing back the Fermi velocity v_F, we get

$$\sigma_T(\omega) = \frac{e^2}{h} \frac{N_T v_F^2}{2\beta k_B T} \log\left(\frac{\gamma k_B T L}{v_F \hbar}\right) \left(\frac{1}{-i\hbar\omega + 0^+}\right)$$

where $v_F k = \gamma k_B T$, $\gamma << 1$ and L is the typical sample size. It may be noted that the spin valley degeneracy factor does not appear in the contribution to the conductivity coming from the turbulence flow. It is useful to compare the conductivity σ_T obtained above with the expression for conductivity due to scattering σ_{sc} which is given by

$$\sigma_{\mathrm{sc}} = \frac{e^2}{h} \frac{N k_B T ln2}{-i\hbar\omega + \kappa k_B T \alpha^2},$$

where α is the electron-electron interaction strength and N denotes the spin valley degeneracy in graphene. Comparing the expressions of σ_T obtained here with that of σ_{sc} obtained without turbulence effects present we see that they have very different dependence on the physical parameters of the system. In particular, $\sigma_T(\omega)$ is independent of the spin-valley degeneracy in graphene. The temperature dependence of σ_T and σ_{sc} are also quite different.

The temperature in our formula is a natural cutoff describing the hydrodynamic region. It represents the background equilibrium state temperature of the graphene sample. The factor N_T present in the expression of the conductance is assumed to be of the same order of magnitude as the uniform electron density in the system. This corresponds to the physical picture of a finite macroscopic fraction of the electron fluid exhibiting turbulent behavior.

The turbulent flow discussed here is a novel consequence of nonlinear electron-electron interactions in graphene. The turbulent flow found has superfluid like properties in the sense that the dispersion relation is linear in the momentum and the system has a macroscopic turbulent flow density. The linear dispersion implies that dissipation through scattering with a defect is only possible when the fluid velocity is greater than v_F. The absence of dissipation implies that the electron flow is frictionless in the turbulent regime.

In order to experimentally access this region, a large graphene sample is required. With a suitable sample, the contribution of the conductivity due to the turbulent flow might be identified through its particular dependence on the system parameters.

2.4 Gamow's Estimate of the Temperature of the Universe

Gamow found an experimental signature for a big bang picture for the formation of the universe by predicting that the current absolute temperature of the universe was ≈ 10. This prediction was ignored and it was only in 1965 that Penzias and Wilson detected a background blackbody temperature. This temperature is now interpreted to be that of a cooling universe.

Gamow's idea was to link an expanding cooling universe to the process of matter formation. In order to do this he supposed that initially the universe was very hot and there was no atoms or nuclei but just electrons, protons neutrons and photons. As the universe cooled the nuclei started to combine and matter started to form. Gamow's idea was that this process of matter formation had to first produce deuterium which has a low binding energy of 2 MeV. The corresponding absolute temperature is $T \approx 10^9$. Gamow's required that the process of deuteron formation should have a fast rate otherwise subsequent steps of matter formation could not happen. He estimated this rate using the measured cross section for deuterium to form by the scattering of a neutron and proton and the density of protons present at that time. The formation rate multiplied by the time gives a number. Gamow assumed this number was of order unity. Thus he set

$$\sigma_T v n t_0 t_0 \approx 1$$

where σ_T is the total cross section for the process of deuterium formation: $p + n \to d + \gamma$, by the collision of a proton p with a neutron n producing a deuterium d and photon γ and v is the relative velocity between the proton and neutron which we taken to be of the order of the thermal velocity. The proton density is $n(t_0)$ where t_0 is the time from the big bang to the time when this process is unidirectional i.e. the deuterium produced is not broken up by energetic photons. Since the binding energy of deuterium is 2 Mev t_0 has to be chosen so that the temperature of the background radiation has energy less than 2 Mev i.e. $t \leq 10^9$ degrees Kelvin. Putting in numbers: $\sigma_T v \approx 10^{-20}$ we see that we must have $n(t_0)t_0 \approx 10^{20}$ where t_0 has to be chosen so the the corresponding temperature of the universe is 10^9 degrees Kelvin. We thus need to relate the time from the big bang to the background temperature of the universe. We also need to determine the time dependence of the proton density function . Both these matters are addressed by two theoretical results which are:

$$\frac{n(t_1)}{T_1^3} = \text{Constant}$$

$$t = \frac{1}{T^2}\sqrt{\frac{3c^2}{32\pi\sigma G}}$$

where σ is the Stefan Boltzmann constant. The first result links the density to the background temperature of the universe, the second links the time to the temperature. The first result follows from two facts The first is that after the process of particle formation stops the number of particles in a given volume of the expanding universe remains unchanged i.e. $n(t)a^3(t)$ is constant where $a(t)$ is the changing length scale. The second is that the entropy S density for a given volume element remains constant for an adiabatically expanding universe. Thus the entropy density $\frac{S}{a^3(t)} = \int \frac{\sigma T^4}{T} = \sigma T^3$. This result implies that $a(t)T$ is constant. From these two results the scale factor $a(t)$ can be eliminated to give the result stated.

$$T_1 = T_0 \left(\frac{n(T_1)}{n(T_0)}\right)^{\frac{1}{3}}$$

We are now ready to get Gamow's result. From the expression relating the time t to the temperature of the universe at that time T we see when $T = 10^9$ degrees Kelvin the corresponding time $t \approx 10^2$ seconds. So that the constraint $n(t_0)t_0 \approx 10^{20}$ tells us that $n(t_0) \approx 10^{18}$. Finally we can measure the current proton density of the universe to get $n(T_1) \approx 10^{-6}$ where $n(T_1)$ is the current density. Substituting these numbers in the formula relating temperature to number density we get

$$\text{Current Temperature } T_1 = 10^9 \times \left(\frac{10^{-6}}{10^{18}}\right)^{\frac{1}{3}} = 10$$

The remaining result to prove is the expression relating the time to the temperature. We give a simple non relativistic argument which gives the required result but can be questioned. Consider a particle of mass m and velocity v at the edge of the universe. It has energy E given by

$$E = \frac{1}{2}mv^2 - \frac{GmM}{r} = k$$

where M is the mass of the universe and k determines the nature of the universe. Here we can take $r = a(t)$. Thus in this situation $k > 0$ describes an expanding universe, $k < 0$ a universe of positive curvature that is closed and $k = 0$ a flat universe. We consider the case of a flat universe. If we now suppose that the universe is radiation dominated then $M = \frac{4c^2 \pi \rho_0 r^3}{r^3}$ where we consider a spherical universe with radiation density $\rho_0 = \sigma T^4$ where the c^2 factor converts the energy density to its equivalent mass. Now since the entropy is taken to constant we must have rT constant. Finally we set $k = 0$ as we are considering a flat universe. Then we have the equation

$$\left(\frac{dr}{dt}\right)^2 = G\frac{4\pi}{3}\frac{\sigma T^4}{c^2}\frac{r^3}{r},$$

from which we get

$$r^2\left(\frac{dr}{dt}\right)^2 = \frac{8G\pi\sigma T^4 r^4}{3c^2}$$

Taking square roots and using the fact that rT is constant gives the result listed. This is an extraordinary example of the power of qualitative estimates to make predictions.

We next describe quantum field theory.

2.5   Quantum Field Theory

Condensed matter physics is the physics of system of many particles that are close together on an atomic scale. Thus we need to use a quantum theory which was set up to describe atomic scale phenomena and we also need to include the fact that systems of interest involve many particles. The fundamental method for doing this is Quantum field theory. Thus we need to understand the methods of quantum field theory.

Let us start by looking at a collection of N identical non interacting particles of mass m and vector momenta p_i that are close together on an atomic scale. There are two very different ways of determining the total energy of such a quantum system.

The first way is to find the energy eigenvalue E_N of the N particle system by solving the Schroedinger's equation:

$$\sum_{i=1}^{N} -\frac{\hbar^2}{2m}\nabla_i^2\Psi(x_1, x_2, ...x_N) = E_N\Psi(x_1, x_2, ..x_N)$$

The second way involves two steps. In the first step Schroedinger equation is solved for a single particle i.e.

$$-\frac{\hbar^2}{2m}\nabla^2\psi(x_i) = E_i\psi(x_i)$$

In the second step the fact that the N particles are identical and non-interacting is used by observing that each particle will have as its energy one of the allowed single particle energy eigenvalues of E_i. Since the particles are non interacting the total energy is then simply

$$E_N = \sum_{i=1}^{N} n_i E_i$$

where n_i is the number of particles among the N that have energy E_i. Thus n_i can range from 0 to N and we must have $\sum_{i=1}^{N} n_i = N$. There seems, at this stage, very little to recommend the second method. It needs us to anyway solve Schroedinger's equation and it seems to be have very limited scope as the total energy of a system of N objects is equal to the sum of the energies of the objects separately only when there are no interactions. But interactions are the heart of physics. In the first approach the inclusion of interactions is easy we simply add a potential energy term to get the modified Schroedinger's equation:

$$[\sum_{i=1}^{N} -\frac{\hbar^2}{2m}\nabla_i^2 + \sum_{i=1}^{N}\sum_{j=1}^{N} V(x_i, x_j)]\Psi(x_1, x_2, ...x_N) = E_N \Psi(x_1, x_2....x_N)$$

where $\nabla_i^2 = \frac{\partial^2}{\partial x_i^2} + \frac{\partial^2}{\partial y_i^2} + \frac{\partial^2}{\partial z_i^2}$ i.e. the operator only differentiates with respect to the vector variable $\mathbf{x_i} = (x_i, y_i, z_i)$. Solving which will give us E_N. Including interactions in the second approach, however, seems to be hopeless as the sum of terms cannot be extended to include interactions.

This problem of introducing interactions in the second approach can be solved, as we now proceed to show, by introducing the ideas of a quantum field theory. Let us explain how this happens. The idea is to regards

$$E_N = \sum_{i=1}^{N} n_i E_i$$

as the eigenvalue of an operator. We have a set E_i of allowed energies for free particles and a set n_i that fix E_N. Once a proper operator associated with this approach is found we will show how an equivalence between this approach and the free Schroedinger equation for N particles can be established. After this step the method of introducing interactions in the second approach can be described. Thus in the first step necessary in this approach is to recognise that the dynamical variables of this approach are the set n_i as different combinations of allowed values from this set give different total energies. The set n_i are a set of integers, hence we need to construct a set of operators which have as eigenvalues which are positive integers or zero. There are a well known set of operators $a_i, a_j^\dagger$ which have this property. These are the energy raising and lowering operators of a harmonic oscillator. We list the commutation rules satisfied by these operators. They are

$$[a_i, a_j^\dagger] = \delta_{i,j}$$

$$[a_i, a_j] = [a_i^\dagger, a_j^*] = 0$$

where $[A, B] = AB - BA$ is called the commutator. The operator $n_i^{op} = a_i^\dagger a_i$ we claim, is exactly the operator we want as its eigenvalues are positive integers or zero. We will establish this result shortly. Assuming this result for the moment, we can construct an operator for the total energy, H_{op}, which satisfies the eigenvalue equation as:

$$H_{op}|n_i> = \sum_{i=1}^{N} n_i|n_i>$$

where $H_{op} = \sum_{i=1}^{N} a_i^\dagger a_i$ and $|n_i> = |n_1, n_2, ..n_i, ... >$ We next establish the following results:

$$a_i^\dagger|n_1, ...n_i, .. > = \sqrt{n_i}|n_1, .., n_i + 1, ... >$$
$$a_i|n_1, ..n_i, ... > = \sqrt{n_i + 1}|n_1, ..n_i - 1, ... >$$

These results tell us that the operator a_i^* raises the value of n_i in a state from n_i to $n_i + 1$ while the operator a_i lowers the value of n_i in a state from n_i to $n_i - 1$. Thus the operator a_i^* increases the contribution to the total energy made by the state labeled by i by an amount E_i while the operator a_i lowers the energy contribution from the state labeled by i by E_i. In colourful language we can say the $a_i^\dagger$ is a creation operator of energy E_i while a_i is a destruction operator of energy E_i. This terminology is very suggestive. It focuses on the particle aspect of the problem and tell us that that quantum operators $a_i, a_j^\dagger$ naturally describe the particle nature of the system. If we recognise that for free particles $E_i = \frac{p_i^2}{2m}$ in momentum space then the label i corresponds to the momentum label p. If we let p be continuous then the operators $a_i, a_j^\dagger$ have continuous labels. Thus these operators are operators labeled by a set of continuous variables p. Objects described by a continuous label are called fields, for example we have the electromagnetic field which gives the value of the electric field at a point in space. Thus our simple non-interacting system of N identical particles has lurking in the background a quantum field theory. It is quantum because operators are present and it is a field theory as the operator degrees of freedom are continuous. In the quantum field theory the single particle momentum p is not an operator it is just a set of three real numbers used to describe the momenta. The new variables that describe the quantum system are the creation and destruction operators. Although the system considered is non-interacting it does have one important property. The approach uses the expression for the energy of a single particle as a function of the particle momentum and describes a state in terms of the variable number of particles of this energy that it contains. The power of this approach is that there is no constraint on what expression for the particle energy one should use. Thus the procedure will work for non-relativistic particles, where the energy $e(p) = \frac{p^2}{2m}$, for relativistic particles, where $e(p) = \sqrt{c^2p^2 + m^2c^4}$ or for photons where $e(p) = c|p|$. One final comment. Since we use creation and destruction operators which have the algebraic properties of the harmonic oscillator it follows that the energy expressions written down must have a zero-point component i.e. the ground state energy has a non-zero energy value even when no particles are present. For the moment we ignore this contribution regarding it simply as defining the zero energy scale but its presence can lead to interesting physical consequences as we have seen.

Let us now establish the properties of the creation and destruction operators stated. There are two ideas we have to use. First we have to use the fact that the physical system is built out of single particles which have a set of fixed allowed energy levels. Second the number of particles which have a specific allowed energy value can change. Thus we need consider three operators: $n_{op} = a^\dagger a, a, a^\dagger$ and use their algebraic properties to derive the results stated. We have suppressed the momentum label p which each of the operators listed carry.

We next state the important lemma:

Calculating Lemma

$$[n_{op}, a] = -a$$

$$[n_{op}, a^\dagger] = +a^\dagger$$

These commutation results are easily derived using the identity $[A, BC] = [A, B]C + B[A, C]$ for operators A, B, C. Next we show that if we have the eigenvalue equation,

$$n_{op}|n> = n|n>$$

where $<n|n> = 1$ then n can only be a positive integer or zero.

Proof

$$[n_{op}, a]|n> = -a|n>$$
$$= (n_{op}a - an_{op})|n>$$
$$(n_{op}a - an)|n> = -a|n>$$
$$n_{op}a|n> = (n-1)a|n>$$

Thus we have established that if $|n>$ is an eigenvector of the operator n_{op} with eigenvalue n then $a|n>$ is also an eigenvector of n_{op} with eigenvalue $(n-1)$ i.e. this eigenvector has a lower eigenvalue. But this process of reducing eigenvalues starting from a fixed eigenvalue n cannot be continue indefinitely as it will lead to n_{op} having negative eigenvalues which we now show is not allowed. Consider

$$n = <n|n_{op}|n> = <n|a^{\dagger}a|n> = <s|s> \geq 0$$

where $a|n> = |s>$. Thus $n \geq 0$ so that for consistency we must set $a|n_{min}> = 0$ to stop the process of lowering eigenvalues. In other words there is a minimum value $n_{min} \geq 0$ below which one cannot go. We set $n_{min} = 0$. Then the action of $a^{\dagger}$ on $|n>$, using the commutation relations and the steps followed for the operator a, can be shown that $a^{\dagger}|n>$ is an eigenvector of n_{op} with eigenvalue $(n+1)$. Thus $a^{\dagger}|0>$ gives an eigenvector of n_{op} with eigenvalue 1. This process of increasing the eigenvalues of n_{op} in steps of one can continue indefinitely. Thus the eigenvalues n can be zero or any positive integer. Our starting eigenvectors $|n>$ were normalised. The action of the operators a, a^* give eigenvectors but these eigenvectors are not normalised to have unit length. We proceed to fix the normalisation factors required. Suppose

$$a|n> = \Gamma_-|n-1>$$
$$a^{\dagger}|n> = \Gamma_+|n+1>$$

To fix the normalisation coefficients Γ_-, Γ_+ we simply work out the norm of the vectors i.e. $a|n>, a^{\dagger}|n>$. We give details of one calculation. Consider

$$||a|n>|| = <n|a^{\dagger}a|n> = n<n|n> = n = |\Gamma_-|^2$$

Thus $|\Gamma_-|^2 = n$. We set $\Gamma_- = \sqrt{n}$ so that we have

$$a|n> = \sqrt{n}|n-1>$$

In a similar way we get

$$a^\dagger|n> = \sqrt{n+1}|n+1>$$

and $n_{op}|n> = a^\dagger a|n> = n|n>$. This is the result we set out to establish. These simple results are very important for quantum field theory calculations.

To summarise we have taken two important steps. The first step was to show that for N non-interacting particles, the total energy of the system can be written in terms of the eigenvalues of a collection of quantum field theory number operators which are built out of creation-destruction operators. The second step was to determine the way these creation destruction operators act on the number labeled eigenvectors introduced.

We now take the third step which is to show how the quantum field theory creation destruction operator approach is equivalent to the N free particle Schroedinger approach. We state the correspondence in the form of two Equivalence Theorems.

Equivalence Theorem 1

Suppose

$$H_{op}|N, E_N> = E_N|N, E_N>$$
$$N_{op}|N, E_N> = N|N, E_N>$$
$$N_{op} = \sum_{i=1}^{N} a_i^\dagger a_i$$
$$H_{op} = \sum_{i=1}^{N} E_i a_i^\dagger a_i$$
$$E_i = \frac{p_i^2}{2m}$$

and there is a state $|0,0>$ such that $H_{op}|0,0> = N_{op}|0,0> = 0$ then the system, described here in terms of momenta labels $i = p_i$ has an equivalent position space representation given by:

$$H_{op} = \int dx \Psi^\dagger(x)(-\frac{\hbar^2}{2m})\nabla^2 \Psi(x)$$
$$N_{op} = \int dx \Psi^\dagger(x)\Psi(x)$$

where $\Psi^\dagger(x), \Psi(y)$ are operators which obey the commutation relations

$$[\Psi^\dagger(x), \Psi(y)] = \delta(x - y)$$

$$[\Psi^\dagger(x), \Psi^\dagger(y)] = [\Psi(x), \Psi(y)] = 0$$

where $\Psi(y) = \int dp e^{-ipy} a_p$, and $\Psi^\dagger(x) = \int dp e^{ipx} a_p^\dagger$ i.e. the position labeled operators $\Psi^\dagger(x), \Psi(y)$ are simply the Fourier transforms of the momentum labeled operators $a_p^\dagger, a_p$. The commutation relations written down follow from those of the creation destruction operators introduced earlier. We can intuitively think of $\Psi^\dagger(x)$ as creating a particle at the point x and $\Psi(x)$ as destroying a particle at the point x.

Let us show how $[\Psi^\dagger(x), \Psi(y)] = \delta(x - y)$ comes from $[a_k^\dagger, a_q] = \delta_{p,q}$. The other results can be proved in the same way. We use Fourier transforms to write

$$\Psi(y) = \int dq e^{-iqy} a_q$$

$$\Psi^\dagger(x) = \int dp e^{ipx} a_p^\dagger$$

Then

$$[\Psi^\dagger(x), \Psi(y)] = \int dq \int dp e^{ipx - iqy} [a_p^\dagger, a_q]$$

Using $[a_p^\dagger, a_q] = \delta_{p,q}$ we get

$$[\Psi^\dagger(x), \Psi(y)] = \int dp e^{ip(x-y)} = \delta(x - y)$$

which established the result.

Next we link the operator approach to the Schroedinger equation approach by defining a function $\Phi(x_1, x_2, ..x_N)$ as follows:

Equivalence Theorem 2

$$\frac{1}{\sqrt{N!}} < 0, 0 | \Psi(x_1) \Psi(x_2) ... \Psi(x_N) | E_N, N > = \Phi(x_1, x_2, ...x_N)$$

We next show that $\Phi(x_1, x_2, ...x_N)$ satisfies Schroedinger's equation. We thus have a bridge connecting the two approaches.

$$\sum_{i=1}^{N} \frac{-h^2}{2m} \nabla_i^2 \Phi(x_1, x_2, ...x_N) = E_N \Phi(x_1, x_2, ..x_N)$$

Proof of Equivalence

The only tool at our disposal in the operator formulation are the commutation relations that are given. Thus in order to derive results from the
operator approach we have to use the commutation relations given. We
proceed to show how this is done. First we establish a simple result. We
note that

$$\frac{1}{\sqrt{N!}} < 0,0|\Psi(x_1)\Psi(x_2)..\Psi(x_N)H_{op}|E_N,N> = E_N\Phi(x_1,x_2,...x_N)$$

This step gives us the right hand side of Schroedinger's equation. To show
that the left hand side has Schroedinger's equation hidden in it we must use
the commutation relations which define the operator approach. The com-
mutation relations appear when include the information that $H_{op}|0,0> = 0$.
This can be added on to the equation written by simply replacing the prod-
uct $< 0,0|\Psi(x_1)\Psi(x_2)...\Psi(x_N)H_{op}|E_N,N>$ by the commutator as follows:

$$< 0,0|\Psi(x_1)\Psi(x_2)..\Psi(x_N)H_{op}|E_N,N>$$

$$= < 0,0|[\Psi(x_1),\Psi(x_2)..\Psi(x_N),H_{op}]|E_N,N>$$

because on the left hand side we have the vector $|0,0>$ and $H_{op}|0,0> = 0$
the term of the commutator with H_{op} on the left hand side gives zero
contribution. We next evaluate the commutator and show that we get
Schroedinger's equation. To do this we need to use the expression for H_{op}
in terms of the operators $\Psi^\dagger(x), \Psi(x)$, and we need the Calculating Lemma:

Calculating Lemma

Given a collection of operators $A_1, A_2, ...A_N, B$ The commutator

$$[A_1A_2....A_N,B] = \sum_{i=1}^{N} A_1...A_{i-1}[A_i,B]A_{i+1}...A_N$$

This result can be proved by induction. First one checks that the result
is true for $N = 2$, then it is assumed to hold for $i - 1$ and from this it is
shown that the result holds for i. Let us proceed to apply this lemma to
our problem. We have

$$[\Psi(x_1)\Psi(x_2)...\Psi(x_N),H_{op}] = \sum_{i=1}^{N} \Psi(x_1)...\Psi(x_{i-1}[\Psi(x_i),H_{op}]\Psi(x_{i+1}..\Psi(x_N)$$

Thus we need to work out the commutator $[\Psi(x_i),H_{op}]$ where $H_{op} = \int dy\Psi^\dagger(y)\frac{-h^2}{2m}\nabla_y^2\Psi(y)$. Using the lemma again we get

$$[\Psi(x_i),\int dy\frac{-h^2}{2m}\nabla_y^2\Psi(y)] = \frac{-h^2}{2m}\nabla_i^2\Psi(x_i)$$

In order to get this result we used the fact that $[\Psi^\dagger(y), \Psi(x_i)] = \delta(y - x_i)$ and all other commutators vanish. Using this result the left hand side of Schroedinger's equation is obtained. Since

$$\sum_{i=1}^{N} < 0,0|\Psi(x_1)\Psi(x_2)..\Psi(x_{i-1}[\Psi(x_i), H_{op}]\Psi(x_{i+1})..\Psi(x_N)|E_N, N >$$

$$= \sum_{i=1}^{N!} \frac{-h^2}{2m}\nabla_i^2 < 0,0|\Psi(x_1)\Psi(x_2)...\Psi(x_i)..\Psi(x_n)|E_N, N >$$

where ∇_i^2 means that only functions of the variable x_i are acted on by the differential operator.

Let us comment on what has been done. The connection bridge suggests a way in which interactions can be included in the quantum field theory approach as we now describe. Furthermore once interactions are introduced properly they give can be used to construct N particle Schroedinger wave-functions. The two approaches are equivalent. We established that the non interacting term modle represents non interacting part of Schroedinger's equation. Let us now interpret this result as follows. The $\Psi^\dagger(x)\Psi(x)$ factor can be regarded as the "particle density operator ", $\rho(x)$. Then this term is simply the "energy " $\frac{-h^2}{2m}\nabla^2$ associated with a single particle. This suggests that interaction energies should be associated with "two- particle densities" that depend on two coordinates x, y and the potential energy $V(|x - y|)$ associated of the two particles. Thus a term of the form:

$$\frac{1}{2}\int dx \int dy \Psi^\dagger(x)\Psi^\dagger(y)V(|x - y|)\Psi(x)\Psi(y)$$

would seem plausible. Once this idea is grasped "three- particle densities" leading to three-body forces and so on can be considered. It is possible to show that the expression for the potential energy contribution written down does, in fact, lead to $\Phi(x_1, x_2, ..x_N)$ satisfying Schroedinger's equation with the potential $V(|x - y|)$. We will not establishing this technical result but this discussion should make it clear that quantum field theory does provide a natural framework for condensed matter physics. The problem is to work out physical properties of a system within this framework.

2.5.0.1 *Fermions*

The presentation of quantum field theory given so far has been for bosons, i.e. particles which have multi-particle wave-functions symmetric under the exchange of particles. This is true for the wave-function

$$\Phi(x_1, x_2, ..x_N) = \frac{1}{\sqrt{N!}} < 0, 0|\Psi(x_1)\Psi(x_2)...\Psi(x_i\Psi(x_{i+1}...\Psi(x_N)|E_n, N >$$

Since the operators $\Psi(x_i)\Psi(x_{i+1})$ commute. To describe fermions we need
to have multiparticle wave functions that are antisymmetric under particle
the exchange of particles. All half integer normal spin systems, such as electrons, are fermions while integer spin systems are bosons. A natural way of
constructing antisymmetric multiple particle wave functions is to introduce
quantum field operators that anticommute. Taking the corresponding N
particle wavefunction to be of the same form as $\Phi(x_1, x_2, ...x_N)$ but with
operators $\Psi(x_i)$ that now anticommute i.e. satisfy the anticommutation
rules:

$$[\Psi^\dagger(x), \Psi(y)]_+ = \delta(x - y)$$

$$[\Psi^\dagger(x), \Psi^\dagger(y)]_+ = [\Psi(x), \Psi(y)]_+ = 0$$

where $[A, B]_+ = AB + BA$ is called the anticommutator, makes
$\Phi(x_1, x_2, ..x_N)$ antisymmetric under particle exchange. It is possible to
show, following the steps described for bosons, that $\Phi(x_1, ...x_N)$ .satis-
fies the N particle Schroedinger's equation. However in order to evaluate
$[H_{op}, \Psi(x_i)]$ the following identity relating commutators to anticommuta-
tors , valid for any three operators A, B and C, has to be used:

$$[AB, C] = A[B, C]_+ - [A, C]_+ B$$

The momentum space anticommutation relations for fermions which follow
from the position space results written down are

$$[a_p^\dagger, a_q]_+ = \delta_{p,q}$$

$$[a_p^\dagger, a_q^\dagger]_+ = [a_p, a_q]_+ = 0$$

An immediate consequence of these anti commutation relations is the important result for the allowed eigenvalues of the number operator $n_p = a_p^\dagger a_p$

$$(n_p)^2 = a_p^\dagger a_p a_p^\dagger a_p = a_p^\dagger(1 - a_p^\dagger a_p)a_p = a_p^\dagger a_p = n_p$$

From this result $(n_p)^2 = n_p$ it follows that the eigenvalues of n_p can only be
either zero or one. Thus only one fermion can occupy a given momentum
state.

2.6 Quasiparticles

Condensed matter systems are always many-body systems which can have unusual highly correlated ground state configurations. For such situations treating the interactions as a small perturbations is not reasonable since a perturbative approach always assumes the free system ground state is a good approximation. Thus system with unusual ground states require non perturbative methods for their analysis. Once the ground state has been constructed the next step is to determine the nature of its excitations. These excitations called "quasiparticles" can be used to determine the physical properties of the system when it is in the non-perturbative ground state or when a many-body system is in a special configuration. Quasiparticles have a restricted range of validity but in this range they can be used to analyse the physical system of interest using a quantum mechanical rather than quantum field theory methods. If however the physical properties of interest are expected to depend on quasiparticle interactions then quantum field theory methods have to be used.

For instance we might be interested in the excitations of a superfluid or a superconductor. In these examples the idea of a quasiparticle is very useful. A quasiparticle describes the excitations of the special ground state constructed. The quasiparticle excitations usually have energy and momenta related in a way that is different from those of the underlying free constituents of the system. Thus the superfluid quasiparticle has energy proportional to the absolute value of its momenta while recently it has been suggested that the quasiparticles associated with novel new materials called topological insulators in the proximity of super conductors are Majorana fermions: objects that are their own antiparticle, predicted theoretically in 1937 but never found before. These Majorana fermions are real, not complex, neutral objects.

A different type of quasiparticle emerges in graphene. In this case the electrons of the system can be replaced by a two dimensional massless Dirac like quasiparticle excitations near a special point of the Brillouin zone of graphene. This quasiparticle is useful for probing the low energy properties of the system.

Let us set up a quantum field theory description for superfluids and superconductors and determine quasiparticles associated with them. The massless Dirac quasiparticle associated with graphene is discussed in a separate chapter.

2.6.1 *Quasiparticles of Superfluid Helium*

The remarkable properties of Superfluid Helium, such as its lack of viscosity, are understood in terms of the phenomenon of Bose–Einstein condensation of the helium atoms. We will see how this happens using ideas of Bogoliubov where the idea of a quasiparticle was first introduced. Our starting point is a quantum field theory Hamiltonian description for a collection of interacting helium atoms. From our discussion of many-body theory we can write down the Hamiltonian as

$$H = H_O + H_I$$
$$H_O = \sum_{\mathbf{k}} E_0 a_{\mathbf{k}}^\dagger a_{\mathbf{k}}$$
$$H_I = \frac{1}{2V} \sum_{\mathbf{k_1},\mathbf{k_2},\mathbf{q}} \mathcal{V}(\mathbf{q}) a_{\mathbf{k_1}+\mathbf{q}}^\dagger a_{\mathbf{k_2}-\mathbf{q}}^\dagger a_{\mathbf{k_1}} a_{\mathbf{k_2}}$$

Let us use the idea of Bose–Einstein condensation. The way we do this is to suppose that there is a superfluid state $|C; N, N_0 >$ of N Helium atoms of which N_0 atoms are zero momentum plane wave states. This description of the superfluid state requires that N_0 is macroscopic. This in turn means that the creation destruction operators acting on this state do not change the state since

$$a_0 |C; N, N_0 > = \sqrt{N_0} |C; N, N_0 - 1 >$$
$$a_0^\dagger |C; N, N_0 > = \sqrt{N_0 + 1} |C; N < N_0 + 1 >$$

For large N_0 we can simply set $a_0 \approx a_0^\dagger \approx \sqrt{N_0}$. Thus the system has a large parameter N_0. The idea of Bogoliubov was to exploit this fact and construct an effective Hamiltonian restricted to act on the assumed superfluid state $|C; N, N_0 >$. This is done by replacing $a_0, a_0^\dagger$, whenever they appear in the Hamiltonian H, by $\sqrt{N_0}$ and only keeping interaction terms which are of order N_0. Carrying out this procedure we finally get

$$H^B = H^0 + H^1$$
$$H^0 = \sum_{\mathbf{k} \neq 0} g(\mathbf{k})$$
$$H^1 = \sum_{\mathbf{K} \neq 0} h(\mathbf{k})[a_{\mathbf{k}}^\dagger a_{-\mathbf{k}}^\dagger + a_{\mathbf{k}} a_{-\mathbf{k}}]$$

where

$$h(\mathbf{k}) = \frac{N_0 \mathcal{V}}{2V}$$
$$g(\mathbf{k}) = \frac{\mathbf{k^2}}{2m} + \frac{N_0 \mathcal{V}}{V}$$

The effective Bogoliubov Hamiltonian H^B is a quadratic function of the creation destruction operators and can be diagonalised. This means that new creation destruction operators $b_{\mathbf{k}}, b_{\mathbf{k}}^{\dagger}$, which satisfy the commutation relations required for creation destruction operators, can be found in terms of which the Bogoliubov Hamiltonian becomes $H^B = \sum_{\mathbf{k}} \mathcal{E}(\mathbf{k}) b_{\mathbf{k}}^{\dagger} b_{\mathbf{k}}$.

The new operators are taken to be

$$b_{\mathbf{k}} = \alpha_{\mathbf{k}} a_{\mathbf{k}} - \beta_{\mathbf{k}} a_{-\mathbf{k}}^{\dagger}$$
$$b_{\mathbf{k}}^{\dagger} = \alpha_{\mathbf{k}} a_{\mathbf{k}}^{\dagger} - \beta_{\mathbf{k}} a_{-\mathbf{k}}$$

where $\alpha_{\mathbf{k}}, \beta_{\mathbf{k}}$ are taken to be real valued functions of $|\mathbf{k}|$ and are called Bogoliubov coefficients. They are fixed by requiring the operators $b_{\mathbf{k}}, b^{\dagger} - \mathbf{k}$ satisfy the commutation relations for creation destruction operators. This leads to the constraint

$$\alpha^2(\mathbf{k}) - \beta^2(\mathbf{k}) = 1$$

Using this result we can write

$$a_{\mathbf{k}} = \alpha(\mathbf{k}) b_{\mathbf{k}} + \beta \mathbf{k} b_{-\mathbf{k}}^{\dagger}$$
$$a^{\dagger} \mathbf{k} = \alpha(\mathbf{k}) b_{\mathbf{k}}^{\dagger} + \beta(\mathbf{k}) b_{-\mathbf{k}}$$

substituting these in the original Hamiltonian, dropping non operator terms, we get

$$H^B = \sum_{\mathbf{k} \neq 0} \mathcal{E}(\mathbf{k}) b_{\mathbf{k}}^{\dagger} b_{\mathbf{k}} + \sum_{\mathbf{k} \neq 0} \mathcal{G}(\mathbf{k}) [b_{\mathbf{k}}^{\dagger} b_{-\mathbf{k}}^{\dagger} + b_{\mathbf{k}} b_{-\mathbf{k}}]$$
$$\mathcal{E}(\mathbf{k}) = [g(\mathbf{k})(\beta^2(\mathbf{k}) + \alpha^2(\mathbf{k})) + 4h(\mathbf{k})\alpha(\mathbf{k})\beta(\mathbf{k})]$$
$$\mathcal{G}(\mathbf{k}) = [g(\mathbf{k})\alpha(\mathbf{k})\beta(\mathbf{k}) + h^2(\mathbf{k})(\alpha^2(\mathbf{k}) + \beta^2(\mathbf{k}))]$$

setting $G(\mathbf{k}) = 0$ and using the result $\alpha^2(\mathbf{k}) - \beta^2(\mathbf{k}) = 1$ obtained earlier in order to get the correct commutation relations for the new operators we can determine the coefficients $\alpha(\mathbf{k}), \beta(\mathbf{k})$ in terms of $g(\mathbf{k})$ and $h(\mathbf{k})$. We get

$$\alpha^2(\mathbf{k}) - \frac{1}{2} = \frac{1}{2} \sqrt{\frac{g2(\mathbf{k})}{g^2(\mathbf{k}) - 4h^2(\mathbf{k})}}$$

$$\beta^2(\mathbf{k}) + \frac{1}{2} = \frac{1}{2} \sqrt{\frac{g^2(\mathbf{k})}{g^2(\mathbf{k}) - 4h^2(\mathbf{k})}}$$

Using these results we finally get

$$H^B = \sum_{\mathbf{k}} \mathcal{E}(\mathbf{k}) b_{\mathbf{k}}^{\dagger} b_{\mathbf{k}}$$

where

$$\mathcal{E}(\mathbf{k}) = \sqrt{g^2(\mathbf{k}) - h^2(\mathbf{k})}$$

$$= \sqrt{\frac{\mathbf{k^2}}{2m}\left(\frac{\mathbf{k^2}}{2m} + \frac{2N_0\mathcal{V}}{V}\right)}$$

The energy $\mathcal{E}(\mathbf{k})$ is called the quasiparticle energy and the operators $b_{\mathbf{k}}^\dagger, b_{\mathbf{k}}$ the quasiparticle creation and destruction operators. By construction they satisfy the creation destruction commutation relations:

$$[b_{\mathbf{k_i}}^\dagger, b_{\mathbf{k_j}}] = \delta_{\mathbf{k_i},\mathbf{k_j}}$$

Let us step back and take stock of what has been done. We started from a Hamiltonian that represented an interacting collection of helium atoms. The next step was to introduced a restriction, namely to postulated that the superfluid state has a macroscopic number N of particles in their zero momentum state. This large parameter N was then used to construct an approximate Hamiltonian. could be approximated once such an assumption was made.

The way this was done was by discarding all interaction terms in the Hamiltonian of order lower than N_0. These steps led to the Bogoliubov Hamiltonian H^B which was quadratic in the creation and destruction operators, which no longer conserved particle number. It had the product of two creation or two destruction operators in it. However H^B did conserve momentum.

The creative idea of Bogoliubov was to show how to introduce new operators which conserved both momentum, particle number and energy. These new operators, he showed, could be constructed as linear combination of the old creation destruction operators. The trick was to combine creation and destruction operators labeled by momentum labels which were $\mathbf{k}$ and $-\mathbf{k}$. When this was done the new operators, by construction, conserved particle number and momentum. In terms of them the Hamiltonian H^B took the simple form:

$$H^B = \sum_{\mathbf{k}} \mathcal{E}(\mathbf{k}) b_{\mathbf{k}}^\dagger b_{\mathbf{k}}$$

The expression for $\mathcal{E}(\mathbf{k}) = \sqrt{g^2(\mathbf{k}) - h^2(\mathbf{k})}$ depended only on the coefficients of H^B. No detailed knowledge of the nature of Fourier transform of the interaction potential $\mathcal{V}$ present in the original Hamiltonian was required. Thus the approach was very general and immediately implied that if any procedure led to a structure similar to H^B it could always be converted to the quasiparticle energy form he had constructed.

The result physically meant that excitations in the superfluid state should have energy $\mathcal{E}(\mathbf{k})$. In particular for $|\mathbf{k}|$ small it was predicted to have the form:

$$\mathcal{E}(\mathbf{k}) \approx \frac{|\mathbf{p}|}{m} \sqrt{\rho \mathcal{V}}$$

where $\rho = \frac{Nm}{V}$. Thus the relation between energy and momentum was drastically modified.

An interesting consequence of this result is that a molecule of mass M_A and speed $|\mathbf{V_A}|$ moving through a system of these low energy quasiparticles could change its energy by scattering off quasiparticles unless its speed was greater than $v_0 = \frac{1}{m}\sqrt{\frac{VN_0m}{V}}$. Thus for $|\mathbf{V_A}| \leq v_0$ the quasiparticles behave as a frictionless superfluid. Let us give the details.

Consider a process in which a molecule of mass M_A and velocity $\mathbf{V}_A$ produces a quasiparticle excitation of momentum $\mathbf{k}$, and changes its momentum to $\mathbf{Q}$ Energy and momentum conservation laws imply that

$$\frac{\mathbf{P}_A^2}{2M_A} = \frac{\mathbf{Q}_A^2}{2M_A} + \mathcal{E}(\mathbf{k})$$
$$\mathbf{P}_A = \mathbf{Q}_A + \mathbf{k}$$

From these it follows that

$$\frac{M_A \mathcal{E}(\mathbf{k})}{|\mathbf{P}|_A |\mathbf{k}|} \leq 1$$

Substituting the low momentum limit of the quasiparticle energy, namely

$$\mathcal{E}(\mathbf{k}) \approx \frac{|\mathbf{k}|}{m}\sqrt{\frac{VN_0m}{V}}$$
$$= \frac{|\mathbf{k}|v_0}{m}$$

we finally get the result stated, namely that

$$v_0 \leq \frac{|\mathbf{P}|_A}{M_A}$$

Thus the quasiparticle energy expression derived can explain the absence of friction that is observed in a superfluid. Our aim is to explain the basic idea. To see more physics applications one of the books on Superfluids listed at the end of this chapter can be consulted.

2.6.2 *Quasiparticles of Superconductivity*

Superconductivity was first observed for mercury where a dramatic reduction of resistance was observed around $10°K$ which was much greater than the slow decrease of resistance expected as the temperature is lowered. Besides a almost vanishing resistance, a superconducting state has another important characteristic. It expels magnetic fields.

We will briefly sketch how these results are understood for a superconductors using the idea of Bose–Einstein condensation. Since the carriers of charge are electrons which, as fermions, cannot collapse down to the same minimum energy state a different approach from that used to model superfluidity was needed. The idea that worked supposed that a pair of electrons, under appropriate conditions, are weakly bound and form boson, called Cooper pair. These Cooper pairs as they were bosons could form a Bose–Einstein condensate. Thus for a superconductor the Cooper pair was the building block of the highly correlated superconducting state and the carrier of supercurrents.

Our task is thus start with a very special Hamiltonian describing interacting electrons and explain how Cooper pairs emerge. On the way we will find that we need to justify on physical grounds the special form for the interaction term that is used. We begin by writing down the effective Hamiltonian used.

$$H = H_0 + H_I$$
$$H_0 = \sum_{\mathbf{k},\sigma} \epsilon(k) a^\dagger_{\mathbf{k},\sigma} a_{\mathbf{k},\sigma}$$
$$H_I = \frac{1}{N} \sum_{\mathbf{k}_1,\mathbf{k}_2} V_{\mathbf{k}_1,\mathbf{k}_2} a^\dagger_{\mathbf{k}_1\uparrow} a^\dagger_{-\mathbf{k}_1,\downarrow} a_{-\mathbf{k}_2,\downarrow} a_{\mathbf{k}_2,\uparrow}$$

The form chosen for the Hamiltonian needs explaining. The free Hamiltonian H_0 is standard. The interacting part is carefully crafted to include a large amount of physics information. This information has to do with the nature of the effective interaction potential $V_{\mathbf{k}_1,\mathbf{k}_2}$. Electrons interact repulsively through the Coulomb interaction. In a solid the electrons also interact with the lattice vibrations, phonons. This interaction gives rise to an attractive force between electrons. A series of calculations suggested that in a narrow low energy region the effective interaction between spin-up, spin-down electrons of opposite momenta could be attractive. This theoretical result has been incorporated in our starting Hamiltonian which has a negative constant $-V_0$ for a small range of values for $(|\mathbf{k}|_1, |\mathbf{k}|_2)$. This effective interaction acts between two spin-up, spin-down electrons carrying opposite momenta. Thus the starting Hamiltonian incorporates in it important physical information.

We are now ready to show how the successful BCS Hamiltonian can be obtained from our starting Hamiltonian. The main step is to introduce an approximations which converts the Hamiltonian from being a quartic function to being a quadratic function of the creation destruction operators. A reduction of a Hamiltonian from a quartic to a quadratic function of creation and destruction operators was also done for the superfluid. There this reduction came from the replacement of the zero momentum creation and destruction operators by a large number $\sqrt{N_0}$ and then dropping terms that were of order lower than N. Here a very different approach, known as the mean field approximation, has to be used. The outcome of the approximation is again to produce, as desired, a Hamiltonian that is a quadratic function of the creation destruction operators, which is then diagonalised by introducing new creation destruction operators (the quasiparticles). The final result of these steps is get a Hamiltonian which is the sum over quasiparticle number operators multiplied by the corresponding quasiparticle energies.

Let us start. The mean field approximation replaces the product of two operators $X.Y$ by

$$XY \approx X < Y > + Y < X > - < X >< Y >$$

The idea of the BCS approach was to take a pair of electrons that form a Cooper pair as the basic operator of the mean field approximation, i.e. to set

$$X = a_{\mathbf{k_1},\uparrow}^{\dagger} a_{-\mathbf{k_1},\downarrow}^{\dagger}$$
$$Y = a_{-\mathbf{k_2},\downarrow} a_{\mathbf{k_2},\uparrow}$$

Making this replacement we get H_{BCS}:

$$H_{BCS} = \sum_{\mathbf{k},\sigma} \epsilon(\mathbf{k}) a_{\mathbf{k},\sigma}^{\dagger} a_{\mathbf{k},\sigma} - \sum_{\mathbf{k}} [\Delta^c a_{-\mathbf{k},\downarrow} a_{\mathbf{k},\uparrow} + \Delta_{\mathbf{k}} a_{\mathbf{k},\uparrow}^{\dagger} a_{-\mathbf{k},\downarrow}^*]$$

where $\Delta_{\mathbf{k}}^c = < X >$ is the complex conjugate of $\Delta_{\mathbf{k}} = < Y >$. The averages are to be taken over the ground state of the Hamiltonian H_{BCS} that we have yet to determine. Now we have unknown functions $< Y >, < X >$ as the parameters of the model. These parameters are determined by imposing a condition of self consistency. We will explain how this done shortly.

The procedure followed gives a Hamiltonian which is quadratic in the creation destruction operators. It does conserve spin and angular momentum but not charge. As in the case of the superfluid these problems are solved by introducing appropriate linear combinations of $a^\dagger, a$ which conserve spin, momentum and charge. These effectively chargeless fermionic excitations are the quasiparticles of the system. Once found they can be used to explore the physics of the superconducting state. The remarkable nature of these quasiparticles is that they are fermions but have zero charge.

The new fermionic operators introduced are:

$$b_{\mathbf{k},\uparrow} = u_{\mathbf{k}}^c a_{,\uparrow} - v_{-\mathbf{k},\downarrow}$$
$$b^\dagger_{-\mathbf{k},\downarrow} = v^c a_{\mathbf{k},\uparrow} + u_{\mathbf{k}} a^*_{-\mathbf{k},\downarrow}$$

The first constraint on the coefficients $u_{\mathbf{k}}, v_{\mathbf{k}}$ comes from requiring the operators $b, b^\dagger$ to satisfy the usual fermion anticommutation relations, for example that $b^\dagger_{\mathbf{k},\uparrow}, b_{\mathbf{k},\uparrow} = 1$. From this we get the constraint

$$|u_{\mathbf{k}}|^2 + |v_{\mathbf{k}}|^2 = 1$$

Once this result is established we can express the old operators in terms of the new operators as

$$a_{\mathbf{k},\uparrow} = u_{\mathbf{k}} b_{\mathbf{k},\uparrow} + v_{\mathbf{k}} b^\dagger_{-\mathbf{k},\downarrow}$$
$$a^\dagger_{-\mathbf{k},\downarrow} = -v_{\mathbf{k}}^c b_{\mathbf{k},\uparrow} + u_{\mathbf{k}}^c b^\dagger_{-\mathbf{k},\downarrow}$$

These expressions can be substituted in H_{BCS} to give:

$$H_{BCS} = \sum_{\mathbf{k},\sigma} E(\mathbf{k}) b^\dagger_{\mathbf{k},\sigma} b_{\mathbf{k},\sigma}$$

where $E(\mathbf{k}) = \sqrt{\epsilon^2(\mathbf{k}) + |\Delta + \mathbf{k}|^2}$ where, as in the case of the superfluid, the coefficient of terms of the type $b^\dagger b^*, bb$ are set equal to zero and the condition for this to happen together with the condition $|u_{\mathbf{k}}|^2 + |v_{\mathbf{k}}|^2 = 1$ is used to show that setting

$$\Delta_{\mathbf{k}} = |\Delta_{\mathbf{k}}| e^{i\theta}$$
$$u_{\mathbf{k}} = |u_{\mathbf{k}}| e^{i\alpha}$$
$$v_{\mathbf{k}} = |v_{\mathbf{k}}| e^{i\beta}$$

By making phase choices we can get the solutions:

$$|u_{\mathbf{k}}|^2 - \frac{1}{2} = \frac{1}{2}\sqrt{\frac{\epsilon(\mathbf{k})}{\sqrt{\epsilon^2(\mathbf{k}) + |\Delta_{\mathbf{k}}|^2}}}$$

$$|v_{\mathbf{k}}|^2 - \frac{1}{2} = -\sqrt{\frac{\epsilon(\mathbf{k})}{\sqrt{\epsilon^2(\mathbf{k}) + |\Delta_{\mathbf{k}}|^2}}}$$

Let us determine the "gap function" $\Delta_{\mathbf{k}}$ introduced as an average over the ground state of the Hamiltonian H_{BCS}. We start from the definition

$$\Delta_{\mathbf{k}} = -\frac{1}{N}\sum_{\mathbf{k}_1} V_{\mathbf{k},\mathbf{k}_1} < a_{\mathbf{K},\downarrow}a_{\mathbf{k}_1,\uparrow} >$$

To work this out we need to replace the operators a by operators b. This step will introduce Δ in the sum as the coefficients relating a,b operators depend on Δ. We also use the results

$$< b^{\dagger}_{\mathbf{k}_1,\uparrow}b_{\mathbf{k}_1,\uparrow} > = n_F(E_{\mathbf{k}})$$

$$< b_{-\mathbf{k}_1,\downarrow}b^{\dagger}_{-\mathbf{k}_1,\downarrow} > = (1 - n_F(E_{\mathbf{k}}))$$

with all other expectation values equal to zero in the BCS state. Thus we end up with an equation for determining $\Delta_{\mathbf{k}}$ which is

$$\Delta_{\mathbf{k}} = -\sum_{\mathbf{k}_1} V_{\mathbf{k},\mathbf{k}_1}\frac{\Delta_{\mathbf{k}_1}}{2E(\mathbf{k}_1)}[1 - 2n_F(E(\mathbf{k}_1))]$$

where $n_F(E)$ represents the Fermi distribution. This is our self consistency condition from which Δ can be determined. It involves thermal averages which we have denoted as $< b^{\dagger}b >$. A short discussion on how such averages can be calculated is given for completeness in the next section where we also discuss how to estimate the function Δ.

We thus have shown how quasiparticles emerge in in two very different situations. The essential step followed in both examples was to convert a starting Hamiltonian, using physical arguments, to a quadratic function of the creation and destruction operators.

The big difference between the superfluid and the superconductor was that the parameters Δ, Δ^c introduced for the superconductor were not directly given by the theory, unlike the parameter N of the superfluid. The superconductor parameters had to be determined, as we saw, by solving an equation. Let us now estimate Δ.

2.6.2.1 *Estimating the parameter Δ*

We estimate Δ for temperature $T = 0$. In order to do this we replace $\Delta_{\mathbf{k}}, V_{\mathbf{k},\mathbf{k}_1}$ by constants Δ_0, V_0, and we can set $n_F = 0$, for $E(\mathbf{k}_1) \geq E_F$. Finally we replace the sum over $\mathbf{k}_1$ by an integral over ϵ. This introduces a density of states factor $D(E_F)$ at $T = 0$. The equation for Δ now becomes

$$1 = V_0 D(E_F)\int_{-\omega_D}^{\omega_D} d\epsilon \frac{1}{\sqrt{\epsilon^2 + \Delta_0^2}}$$

$$1 = V_0 D(E_F)\text{arcsinh}(\frac{\omega_D}{\Delta_0})$$

where $D(E_F)$ is the density of states and the integration limits are $\pm\omega_D$, the Debye frequency and the sign of $V_0 \leq 0$, meaning it is an attractive potential, has been included. It is in this range that the potential is supposed to be non-zero. This gives

$$\Delta_0 = 2\omega_D e^{-[\frac{1}{V_0 D(E_F)}]}$$

This result clearly shows that the process of condensation is non-perturbative as the dependence of Δ on V_0 is not linear in V_0 even for $V_0 \to 0$.

2.6.3 *Thermal Averages*

For estimating Δ we used the following result for the thermal average $< b_{\mathbf{k},\uparrow}^{\dagger} b_{\mathbf{k},\uparrow} > = n_F(E(\mathbf{k}))$ Let us briefly sketch how this result can be obtained using the methods of creation destruction operators. The partition function Z for the grand canonical ensemble in operator form is $Z = \text{Trace}(\rho(\beta))$ where

$$\rho(\beta) = e^{-\beta(E-E_F)} = e^{-\beta \sum_{\mathbf{k}_1}(E(\mathbf{k}_1)-\mu)b_{\mathbf{k},\uparrow}^{\dagger} b_{\mathbf{k},\uparrow}}$$

where $\beta = \frac{1}{k_B T}$, k_B is the Boltzmann constant and T the temperature, $E = \sum_{\mathbf{k}} b_{\mathbf{k}}^{\dagger} b_{\mathbf{k}}$, $E_F = \mu \sum_{\mathbf{k}} b_{\mathbf{k}}^{\dagger} b_{\mathbf{k}}$ and μ the chemical potential, we suppress the spin descriptions $\uparrow$ in the expressions for E, E_F and will continue to do so. The thermal average $< b_{\mathbf{k}}^{\dagger} b_{\mathbf{k}} >$ is defined to be

$$< b_{\mathbf{k}}^{*} b_{\mathbf{k}} > = \frac{\text{Trace } [e^{-\beta(E-E_F)} \, b_{\mathbf{k}}^{\dagger} b_{\mathbf{k}}]}{\text{Trace } [e^{-\beta(E-E_F)}]}$$

where E_F is the fermi energy. Let us write $< b_{\mathbf{k}}^{\dagger} b_{\mathbf{k}} > = n_F(E(\mathbf{k}))$. We also note that since the operators $b_{\mathbf{k}_i}^{*}, b_{\mathbf{k}_i}$ with different $\mathbf{k}_i$ labels commute the sum of exponentials can be written as a product of exponentials and as a result we have

$$n_F(E(\mathbf{k})) = \frac{\text{Trace } e^{(E(\mathbf{k})-\mu)b_{\mathbf{k}}^{\dagger} b_{\mathbf{k}}}}{\text{Trace } e^{-\beta(E(\mathbf{k})-\mu)}}$$

To evaluate n_F we need to use the anticommutation relations $b^{\dagger}, b = 1$, the cyclic property of the trace and establish a switching lemma which tells us what happens when the operator $\rho(\beta)$ and say the operator switch places.

Switching Lemma

$$\rho(\beta)b^\dagger \to e^{-\beta\epsilon)}b^\dagger \rho(\beta)$$

where $\epsilon = E(\mathbf{k}) - \mu$. Using this we can establish the result

$$n_F(E(\mathbf{k})) = \frac{1}{e^{\beta\epsilon(\mathbf{k})} + 1}$$

The steps to be followed are

$$\text{Trace } \rho(\beta)b^\dagger b \to e^{-\beta\epsilon}b^\dagger \rho(\beta)b$$
$$\to \text{Trace } e^{-\beta\epsilon}bb^\dagger \rho(\beta)$$
$$\to \text{Trace } e^{-\beta\epsilon}[-b^\dagger b + 1]\rho(\beta)$$

where we first used the switch, then the cyclic property of the trace and finally used the anticommutation relations of the $b^\dagger, b$ operators.

Some Consequences of the BCS Model

The two main features of superconductivity we stated were that they have zero resistance and that they expel magnetic fields. Both of these features can be understood from a simple set of assumptions made by London. We will explain London's approach and then show how they can be understood from the BCS model.

London assumed a two component model for a superconductor. The electron current had a normal component where resistance was present. Resistance of a current comes from the scattering of electrons as they moves through a medium. Three forms of electron scattering have been identified. They are electron- electron scattering, electron-phonon scattering and a special type of scattering where a electron loses its momentum by scattering with the lattice itself. A way to describe these different contributions to resistance is to associate relaxation times to each of the processes described. The presence of a current component with no resistance means that such electrons do not scatter. In this case we should have

$$\mathbf{j}_s = -en_s\mathbf{v}_s$$
$$\frac{d\mathbf{v}_s}{dt} = -e\frac{\mathbf{E}}{m}$$

where $\mathbf{j}_s$ was the supercurrent, $\mathbf{v}_s$ its velocity controlled by the external field $\mathbf{E}$ and m the electron mass. Some immediate consequences follow from these equations.

$$\frac{\partial \mathbf{j}_s}{\partial t} = e^2 \frac{n_s \mathbf{E}}{m}$$

$$\frac{\partial (\nabla \times \mathbf{j}_s)}{\partial t} = -\frac{e^2 n_s}{m} \nabla \times \mathbf{E}$$

$$= -\frac{e^2 n_s}{mc} \frac{\partial \mathbf{B}}{\partial t}$$

where Maxwell's equation is used and $\mathbf{B}$ is the magnetic field. From this it follows that

$$\nabla \times \mathbf{j}_s = -\frac{e^2 n_s}{mc} \mathbf{B} + \mathbf{C}(\mathbf{r})$$

Replacing $\mathbf{B}$ by $\nabla \times \mathbf{A}$ where $\mathbf{A}$ is the vector potential we get

$$\mathbf{j}_s = -\frac{e^2 n_s}{mc} \mathbf{A}$$

This is a key result of London which followed from the assumption that the flow of electrons were not hampered by scattering. Such an assumption implied unbounded speed of flow. Setting aside this worry for the moment we see the conclusion also required setting $\mathbf{C}(\mathbf{r}) = 0$ which is a boundary condition statement that needs physical justification. We also need to include constraints that follow from the requirement of charge conservation. Let us start with the charge conservation constraint. Charge conservation requires the four vector condition $\frac{\partial j_\mu}{\partial x_\mu} = 0$, be satisfied. Imposing this condition on the supercurrent, which is proportional to the vector potential, requires that $\nabla . \mathbf{A} = 0$ (London gauge). Setting $\mathbf{C}(\mathbf{r}) = 0$ is justified because the "rigidity" of the BCS ground state does not allow changes to happen when the $\mathbf{B}$ is introduced. The presence of a gap means that changes can only happen when there is sufficient energy to overcome this barrier.

Once these points of self consistency are noted we can move on to discuss an important consequence of the London equation, namely that it implies that magnetic fields are expelled from a superconducting region. This is the Meissner effect. To see how this result follows from the London equation we use Ampere's law and the gauge condition necessary for charge conservation, $\nabla . \mathbf{A} = 0$

$$\nabla \times \mathbf{B} = \frac{4\pi}{c} [\mathbf{j}_s + \mathbf{j}_n]$$

where $\mathbf{j}_n$ is the normal current. Thus we get,

$$\nabla^2 \mathbf{B} = -\frac{4\pi e^2 n_s}{mc^2} \mathbf{B}$$

This equation tells us that a magnetic field $\mathbf{B}$ is expelled from a region where n_s is non zero and thus explains the Meissner effect.

We next turn to a very useful set of equations for studying the physics of non homogeneous superconductors, known as the Bogoliubov–de Gennes (BdG) equations.

2.7 The Bogoliubov–de Gennes Equations

Homogeneous superconductivity are those where the $\mathbf{k}$ dependence of the term $\Delta_{\mathbf{k}}$ can be neglected. Such an approximation helps us to understand the basic features of superconductivity rather easily. However non homogeneous superconductors exist. A simple way of modeling such systems is by using the Bogoliubov–de Gennes equations (BdG). These equations can also be used to incorporate "proximity" effects. We will use this feature of the BdG equations to briefly discuss Majorana fermions which were originally introduced as a way to describe neutrinos (mass zero fermions) but now seem to exist in the vortex of a superconducting layers next to a topological insulators.

The starting point for setting up the BdG equations is the BCS Hamiltonian

$$H_{BCS} = \sum_{\mathbf{k}} [\epsilon(\mathbf{k}) a^{\dagger}_{\mathbf{k},\sigma} a_{\mathbf{k},\sigma} - \Delta^{c}_{\mathbf{k}} a_{-\mathbf{k},\downarrow} a_{\mathbf{k},\uparrow} - \Delta_{\mathbf{k}} a^{\dagger}_{\mathbf{k},\uparrow} a^{\dagger}_{\mathbf{k},\downarrow}] + E_{BCS}$$

where E_{BCS} is the BCS ground state energy. From this Hamiltonian the following commutation relations between H_{BCS} and the creation destruction operators $a^{\dagger}, a$, can be worked out, using fermion anticommutation relations, of $a^{\dagger}, a$ and the following useful identity: $[AB, C] = ABC - CAB = AB, C - A, CB$, where $[X, Y] = XY - YX$ represents a commutator and $X, Y = XY + YX$ represents an anticommutator. We have

$$[H_{BCS}, a^{\dagger}_{\mathbf{k},\uparrow}] = \epsilon(\mathbf{k}) a^{\dagger}_{\mathbf{k},\uparrow} - \Delta^{c}_{\mathbf{k}} a_{-\mathbf{k},\downarrow}$$

$$[H_{BCS}, a_{-\mathbf{k},\downarrow}] = -\epsilon(\mathbf{k}) a_{-\mathbf{k},\downarrow} - \Delta_{\mathbf{k}} a^{\dagger}_{\mathbf{k},\uparrow}$$

Let us introduce two excitations, namely

$$|+\mathbf{k}, 1> = a^{\dagger}_{\mathbf{k},\uparrow} |BCS>$$
$$|-\mathbf{k}, 2> = a_{-\mathbf{k},\downarrow} |BCS>$$

where $H_{BCS}|BCS> = E_{BCS}|BCS>$ Then we get

$$H_{BCS}|+\mathbf{k}, 1> = (E_{BCS} + \epsilon(+\mathbf{k}))|\mathbf{k}, 1> -\Delta^{c}_{\mathbf{k}}|-\mathbf{k}, 2>$$
$$H_{BCS}|-\mathbf{k}, 2> = (E_{BCS} - \epsilon(+\mathbf{k}))|-\mathbf{k}, 2> -\Delta_{\mathbf{k}}|+\mathbf{k}, 1>$$

Let us consider the structure of $H_{BCS} - E_{BCS} = H$. We note that H is a 2 by 2 matrix, in the basis states introduced, with a mixing term between states that comes from pairing term Δ.

In the Bogoliubov–de Gennes approximation this 2 by 2 structure of H is retained but the detailed nature of the two operators H_0 and Δ that come from the BCS theory is ignored. Instead H_{BdG} is written in real space with operator H_0 replaced by a one particle Schroedinger operator with a "chemical" potential μ , which could be temperature dependent, and a "pairing" potential $\Delta(\mathbf{r})$. The states labeled by $\mathbf{k}$ are now labeled by $\mathbf{r}$. The pair of equation that result from these replacements are the Bogoliubov–de Gennes equations. They are very useful for non homogeneous superconductors.

Thus in the Bogoliubov–de Gennes approximation we set the operator $H_0(\mathbf{r}) = \frac{-\hbar^2}{2m}\nabla^2 - \mu + V(\mathbf{r})$ a single particle operator, and $\Delta(\mathbf{r})$ is assumed to be a given function of $\mathbf{r}$ which encourages pairing. This effective single particle description where both $V(\mathbf{r})$, and $\Delta(\mathbf{r})$ need not have lattice periodicity is an enormous simplification and leads to physical predictions that agree with experimental results.

In matrix form we can write the operator H_{BdG} as

$$H_{BdG} = \begin{pmatrix} H_0(\mathbf{r}) & -\Delta(\mathbf{r}) \\ -\Delta(\mathbf{r}) & -H_0(\mathbf{r}) \end{pmatrix}$$

The corresponding basis states are now $\psi_1(\mathbf{r}, \uparrow), \psi_2(\mathbf{r}, \downarrow)$. A very interesting application of the Bogoliubov–de Gennes equations is to a topological insulators that are placed on a superconductor. The presence of boundaries makes the superconductor non homogeneous. The surface of a topological insulator is known to have gap-less conducting points, where electrons can be represented by a two dimensional mass less Dirac equation. This point is called a Dirac point. In our section on topology and K theory we explained how this happens. When a superconductor is placed on such a surface the electrons of the topological insulator are influenced by the Cooper pairs of the superconductor. A way of modeling this interaction is called the proximity approximation and the BdG equations. Using this approach we now show how the topological insulator-superconductor system has, on the superconductor side, massless neutral fermions that are their own antiparticle, first suggested as a possible description of neutrinos in 1934 by Ettore Majorana. It is remarkable that such an exotic object appears as a quasiparticle in condensed matter physics. These exotic excitations are of interest because they obey non-conventional statistics and could be useful for constructing quantum computers. We proceed to construct an example of a Majorana fermion by first discussing a symmetry between particles and antiparticles present in the Dirac equation, we then show how topological arguments gives zero energy fermions and combine these ideas to construct a two dimensional model which has Majorana fermion excitations.

Majorana fermions are their own antiparticle they also have charge conjugation symmetry which is a symmetry that exists between particles and their antiparticles. We start by describing features of this symmetry that we need.

Charge Conjugation Symmetry and Majorana Fermions

The Dirac equation contains both particles and antiparticles. In condensed matter the antiparticles correspond to holes. The Dirac theory is symmetric under particle-antiparticle exchange. This symmetry is called charge conjugation symmetry. The symmetry operation replaces a particle by an antiparticle which requires that a sign change of charges but it keeps the space-time variables such as mass, momentum and spin unchanged. In this regard this symmetry is very different from time reversal and parity symmetry: it has two distinct particles with different charges. The degrees of freedom of the Dirac system are the sum of the degrees of freedom for the particles and and those of the antiparticle. For a neutral fermion the particle and the antiparticle could be the same. If this the case we have Majorana fermions.

We restrict our discussion of charge conjugation symmetry to the two dimensional Dirac equation which we write as:

$$[\gamma_\mu(\partial_\mu - ieA_\mu) + m]\psi = 0$$

where ψ, A_μ are field operators and e is the electron charge. The two dimensionality of the problem means that the index $\mu = 0, 1, 2$ and $\gamma_0 = \sigma_3, \gamma_1 = i\sigma_2, \gamma_2 = -i\sigma_1$, where $\sigma_i, i = 1, 2, 3$ are the Pauli spin matrices. This is a choice of representing the gamma matrices. Other choices are possible. We now suppose there is an operator C that transforms a particle into an antiparticle with the same mass and spin i.e. we require that $C\psi C^{-1}$ is the field operator for the positron namely,

$$[\gamma_\mu(\partial_\mu + ieA_\mu) + m]C\psi C^{-1} = 0$$

where the sign change for the charge in the equation has changed. This implies

$$CA_\mu C^{-1} = -A_\mu$$
$$C\psi_\alpha C^{-1} = C_{\alpha\beta}\overline{\psi}_\beta$$

where $\overline{\psi} = \psi^\dagger \gamma_0$, $\psi^\dagger$ the adjoint of ψ and $\overline{\psi}$ satisfies the equation,

$$\overline{\psi}[\gamma_\mu^T(\partial_\mu + ieA_\mu) - m] = 0$$

and γ_μ^T is the transpose of γ_μ. Comparing this equation with the equation that $C\psi C^{-1}$ satisfies we get the conditions

$$C^{-1}\gamma_\mu C = -\gamma_\mu^T$$
$$C^T C^{-1} = \text{a c-number}$$

This gives $C = i\gamma_2$ and $C\psi_\alpha C^{-1} = (i\gamma_2)_{\alpha\beta}\overline{\psi}_\beta$. The condition for a Majorana fermion can now be stated. It is

$$\psi_\alpha = (i\gamma_2)_{\alpha\beta}\overline{\psi}_\beta$$

Let us write out this condition in greater detail. Let (u, v) be the components of ψ with u the particle wavefunction and v the antiparticle wavefunction, then $(u^*, -v^*)$ are the components of $\overline{\psi}$. Imposing the Majorana condition relates the particle wavefunction to the antiparticle wavefunction, namely it requires that $v = -u^*$ i.e. the particle and antiparticle wavefunctions are the same modulo a phase. This form of the Majorana condition depends on the representation of the gamma matrices that is used. There is a particular representation for which the wavefunctions are all real and the connecting matrix for charge conjugation is the identity matrix. Thus the Majorana fermion has half the degrees of freedom of a Dirac fermion.

The next idea we need is the way a (mid gap) mass less fermionic excitation can be constructed.

2.8   Topology and Fermion Zero Energy Modes

Jackiw and Rebbi made a startling discovery by showing that one dimensional fermions in a background potential of a topological kink could have a zero energy localised fermion with fractional charge. Subsequently Jackiw and Rebbi extended the idea to two dimensional fermions in the background of topological vortices. Both these theoretical predictions have been observed in condensed matter systems. Both results show the power of topology to predict new excitations. The result of Jackiw and Rossi that two dimensional fermionic solitons, of zero energy, localised at the core of a

topological vortex, can form is used to show that such excitations can be
Majorana fermions.

We first summarise the result of Jackiw and Rebbi for a one dimensional
Dirac particle interacting with a scalar potential which is a topological kink
and write down an associated zero energy fermionic solution. The Dirac
equation is

$$i\partial_t\psi(z,t) = (\alpha p_z + \beta\frac{m}{\kappa}\phi(z))\psi(z,t)$$
$$\phi(z) = \kappa\tanh(\lambda z)$$

and we choose $\alpha = -\sigma_3, \beta = \sigma_2$ where $\sigma_i, i = 2,3$ are Pauli spin matrices.
The potential $\phi(z)$ is zero for $z = 0$ and interpolates between $=\kappa$ for $\phi\pm\infty$.
This equation has a zero energy fermionic solution which is,

$$\psi_0 = e^{-\frac{m}{\lambda}\ln(cosh\lambda z)}\chi$$

where χ satisfies the condition $-i\chi = \alpha\beta\chi$ Such a solution is called a
mid gap mode in condensed matter. A realisation of this excitation has
been found in polyacetylene. The solution shows the power of topology to
produce unexpected effects. The kink introduced represents a domain wall.

Next we move to the case of a two dimensional Dirac equation interacting with a topological vortex and again write down its associated zero
energy mode (mid gap mode) equation describing a fermionic excitation.
The interaction of the fermion with the superconducting vortex is done by
using the BdG equations. Such a way of introducing interactions between
a topological insulator and on a superconductor is called the proximity
approximation. We then write down a zero energy excitation for which
the particle and antiparticle are the same. This establishes the theoretical
possibility of a Majorana fermion to exist in a condensed matter system.

2.8.1 *The Proximity Approximation and Majorana Fermions*

Topological insulators have time reversal invariance and have a mass less
surface Dirac excitation. The Dirac excitation of the topological insulator
thus must be constructed to be time reversal invariant. In order to do
this we consider a two dimensional Dirac system built out of a related
pair of wavefunctions $\Psi = (\psi, i\sigma_2\psi)$. The next step is to introduce an
interaction of these Dirac excitations with the superconductor. This is done
by using the Bogoliubov–de Gennes equation with the pairing interaction

potential selected to be a s-wave vortex excitation of the superconductor as a "proximity" effect. The proximity Hamiltonian $\mathcal{H}$ is then,

$$\begin{pmatrix} \sigma\cdot\mathbf{p} + \sigma_z h - \mu & \Delta \\ \Delta^* & -\sigma\cdot\mathbf{p} + \sigma_z h + \mu \end{pmatrix}$$

where $\mathbf{p} = (p_x, p_y)$, h is a Zeeman magnetic field and μ the chemical potential. The eigenvalue problem for $\mathcal{H}$ has the following components

$$\begin{pmatrix} u_1 \\ u_2 \\ v_1 \\ v_2 \end{pmatrix}$$

This system with $\Delta(x,y) = |\Delta(r)|e^{in\theta}$ in polar coordinated is a vortex with winding number n. Such a "potential" term leads to topological consequences. The operator $\mathcal{H}$ has an index which is,

$$\text{index} = \int dl_i \epsilon_{ab} \overline{\Delta}_a \partial_i \overline{\Delta}_b = n$$

where $\overline{\Delta}_a = \frac{\Delta_a}{\sqrt{\Delta_1^2 + \Delta_2^2}}$, and the line integral is taken at spatial infinity. However in the presence of h, μ the index theorem gets modified as these terms lead to two zero energy states when the winding number n bigger than two, forming a pair of non zero energy states. Thus for an even number n the index is zero while for an odd number it reduces to $n = 1$.

An explicit fermionic zero energy state can be constructed. It is

$$u_1 = \sqrt{\mu + h} J_l(r\sqrt{\mu^2 - h^2}) e^{il\theta - \int^r ds[\Delta(s)]}$$

$$u_2 = \sqrt{\mu - h} J_{l+1}(r\sqrt{\mu^2 - h^2}) e^{il\theta - \int^r ds[\Delta(s)]}$$

where $l = \frac{n-1}{2}$. We have dropped some phase factors. This solution requires that $v_1 = -u_1^*, v_2 = u_2^*$ i.e. these zero mode solutions, localised at the core of the vortex Δ, with half the degrees of freedom of a Dirac particle, are Majorana fermions. There is a condition on the parameters that has to be satisfied in order to make the solution normalisable. It is

$$\mu^2 + |\Delta(\infty)|^2 > h^2$$

We can easily check that the solution constructed has charge zero. The charge density for an electron is $-e|u|^2$, for the positron it is $+e|v|^2$. Thus for our solution the charge density is: $+e[|v_1|^2 + |v_2|^2 - |u_1|^2 - |u_2|^2] = 0$.

Our example demonstrates a theoretical possibility. However realistic physical models of a topological insulator in proximity to a superconductor have been constructed that predict Majorana fermion quasiparticles with non standard statistics. There is also experimental evidence for their existence.

Further Reading and Selected References

E.T. Whittaker, History of Aether and Electricity Vol 2, (Longman Green, 1910) gives the history of the birth of quantum theory and quantum mechanics.

C. Itzykson and J.B. Zuber, Quantum Field Theory (Dover,1980) discusses the Lamb shift in terms of fluctuations due to zero point energy and provides a thorough discussion of relativistic quantum field theory.

K. Milton, The Casimir Energy (World Scientific, 2001) discusses the Casimir Effect and related matters thoroughly with many references.

G. Preparata, QED Coherence in Matter, World Scientific (1995) discusses spontaneous coherent structure formation from a different perspective.

X. Zhang, H. Lhuissier, Chao Sun and D. Lohse, Phys. Rev. Lett. 112, 144503(2014) discusses surface nanobubbles from a classical perspective and gives references to other experimental work.

G. Mahan, Many Particle Physics (Kluwer, 2003). This book gives a good introduction to standard Green function methods of many body theory.

I. Sachs, S. Sen and J. Sexton, Elements of Statistical Mechanics (Cambridge University Press, 2006) includes a short introduction to many body theory and temperature field theory.

M. Tinkham, Introduction to Superconductivity (McGraw Hill) is a good introduction to the BCS theory.

A.B. Migdal and V.P. Krainov, Approximation Methods in Quantum Mechanics, W.A. Benjamin, Inc (1969) contains many examples of qualitative arguments.

A.B. Migdal, Nuclear Theory: the Quasiparticle Method W.A. Benjamin, Inc (1968), discusses general properties of fermi systems using quasiparticles.

M.I. Monastyrsky (Ed), Springer (2006), Topology in Condensed Matter. Contains a collection of articles illustrating the wide range of systems where topological methods have been used.

R. Feynman, Lectures in Physics, Vol3, Addison-Wesley (1967), discusses superconductivity.

V.E. Zakharov, V.S. L'vov and G. Falkovich, Kolmogorov Spectra of Turbulence, Springer-Verlag, Berlin (1992), discusses weak wave turbulence with many examples.

M. Rakowski and S. Sen, Phys. Rev. E 53, 586 (1996) discusses quantum weak wave turbulence.

K.S. Gupta and S. Sen, Europhys. Lett. 90, 34003 (2010), discusses turbulence in graphene.

R. Jackiw, Diverse Topics in Theoretical and Mathematical Physics (World Scientific, 2014). Very clear and full of ideas many, pioneered by Jackiw.

C. Rebbi and R. Jackiw, Phys. Rev. D 13, 3398 (1976). Example of a one dimensional fermionic solitons created due to topology.

Rossi and R. Jackiw, Nucl. Phys. B 190, 681 (1981). Example of a two dimensional fermionic solitons localised at a vortex core.

Chamon et al., Phys. Rev. B 81, 224515. Model of Majorana fermions described in this chapter. Very clear and readable.

Chapter 3

Topology and Geometry

In this chapter we discuss topological ideas and methods. There is now growing awareness that such methods are useful for understanding and predicting the behaviour of condensed matter systems.

Topology is the abstract study of continuity. It introduces mathematical structures which remain unchanged under continuous deformation and is often described to be "rubber sheet geometry" as it focuses on features of space that do remain unchanged under continuous deformations. Let us list a few examples where topological methods are useful.

(1) Is our planetary system stable? To answer this question we do not need to determine the position of all the planets but only to know that the orbit of all the planets remain in a bounded region. In fact Poincare invented topological methods to tackle this problem.

(2) Are there gapless states? Here the interest is not on the precise details of the energy spectrum of a system but on whether the Hamiltonian describing the system has zero energy modes. Sometimes zero modes can be found using symmetry arguments. There are also powerful topological method for determining if they are present as we will show later on.

(3) Are there localised excitations possible in a system (solitons)? Again there are topological ways of finding out if localised, stable excitations exist for a given system. Here homotopy groups of topology play a role.

(4) Can one understand the nature of possible defects allowed? Defects in an ordered medium can be classified using Homotopy groups of topology.

(5) If a crystal symmetry group changes from a higher symmetry to a subgroup symmetry are there selection rules forbidding certain subgroup symmetries? Here the topological method of Morse theory can help.

(6) Can large quantum coherent systems be created? This is a problem of great interest. A interesting approach is to construct topologically protected quantum entangled ground states i.e. systems whose coherence can only be destroyed by non continuous deformations.

This list is not comprehensive but gives a flavour of the wide variety of situations where topological ideas show up.

Let us try to explain how "rubber sheet geometry" can be constructed. Our starting point is to recall the way geometry is defined.

Many types of geometries are possible such as Hyberbolic geometry, Riemannian geometry, Kahler geometry to name a few. Each geometry defines a space where there are points and a notion of distance between them.

For example, in Euclidean geometry the distance between points is defined by the Pythagoras theorem. The distance $d(x, y)$ between two points (x, y) does not change if the cartesian coordinates used are rotated or if the origin of the coordinate system is translated. It is invariant under rotations and translations.

With a notion of distance in place the intuitive concept of continuity can be defined which leads on to the idea of a limit which leads on to calculus. Let us recall how this is done. Continuity of a real valued function $f(x)$ at a point x_0 tries to capture the intuitive idea that if the distance between two points $f(x), f(y)$ is small then they come from two close points (x, y). The precise definition of continuity is that for any choice $d(f(x), f(x_0)) < \epsilon$ (an open interval) we can find a δ such that $d(x, x_0) < \delta$ (a different open interval). In words no matter how close we choose two points $f(x), f(x_0)$ to be we can always find a pair of points x, x_0 the distance between which is less than a δ, where δ depends on the choice made for ϵ. Using continuity the idea of a limit of a function as a statement that if the distance $d(f(x), f(x_0)) \to 0$ as $d(x, x_0) \to 0$ then $f(x_0)$ is the limit of $f(x)$ as $x \to x_0$. Taking the limit of a function introduces a new operation of mathematics that acts on functions. It leads on to calculus. Thus continuity, distance and limits are important ideas needed to introduce calculus. How can the idea of continuity be introduced in "rubber sheet geometry" where there is no well defined notion of distance?

This is the problem that we have to solve if we want to introduce topology as a useful form of "rubber sheet" geometry. The first step is to find a way of describing continuity without using the idea of distance as distances change under deformation. Once this step is taken our next task will be to construct a geometry which is "invariant under continuous deformations". To do this we need a new idea.

The definition of continuity we gave used two ideas. The first idea was that of distance. The second idea was that of an open interval. An open interval was defined by the inequality $d(x, x_0) < \delta$ which determined the set of all points x that belonged to the open set. Intuitively it seems reasonable to suppose that the open interval remains open under continuous deformations. This suggests that we rephrase the statement of continuity given in terms of the open intervals. Such a formulation is possible. It tells us that a function f which maps an open O set to another open set V is continuous if the inverse image of V under f is an open set of O. In this rephrase only open sets are used and no notion of distance used although the open sets themselves were defined using the idea of distance. Open sets can thus be used to define continuity of functions and they are invariant under continuous deformations and are thus natural objects to use in topology. Our next step is to use the open sets as building blocks for a new type of geometry. This would be "rubber sheet" geometry.

Euclidean geometry, we pointed out, is defined in terms of a set of points and a distance function between them. The set of points is the undefined fundamental objects of the subject. For topology the analogue of the set of point is, as we have suggested, the set or collection of open sets and we now propose that the analogue of the distance function satisfying certain rules, will be a set of rules that open sets must obey. The precise conditions required for open sets to give a "topological space" will be described shortly. The important point to grasp is that the collection of open sets in topology is not defined by using a distance function but is the fundamental object of the subject. The aim of the subject is to find and study objects that are invariant under continuous deformations. These invariants, studied in topology, are found to be important for physics.

Our next step is to introduce useful mathematical ideas that we need and also give a precise definition of a topological space. The ideas we consider are:

(1) The idea of a manifold.
(2) The idea of differential forms.
(3) The idea of homotopy, homology and cohomology.
(4) The idea of Fibre Bundles and Vector Bundles.

Let us explain what these ideas are and why they are useful. There are many situations where Euclidean distance is not appropriate. For instance if we want to describe the distance between two points on curved surface such as a sphere. However two points on the surface of a sphere that are close together do have a distance which can be approximated by the Euclidean distance. Spaces that have the property that the distance between points close together can be approximated by the Euclidean distance are called manifolds. For manifolds a scheme of representing them by patches of Euclidean space is possible. By representing a patch of a manifold by Euclidean space we mean that an invertible map taking points on a patch of the manifold to Euclidean space exists. If the map is smooth then this procedure can be used to introduce calculus on a manifold. The idea is to carry out usual calculus operations (differentiation and integration) in the Euclidean space associated with a patch of the manifold and then use the invertible map to transfer the result to the manifold. This is a conceptual step. In many situations the precise nature of the map is not important. Many different maps relating a manifold to Euclidean space are possible. Each map provides Euclidean coordinates for the manifold. If two patches overlap then points of the manifold in the overlap region belonging to two different patches will have two different coordinate representations and formula for relating the two sets of coordinates needs to be given.

Thus calculations on the manifold are coordinate dependent. For this reason mathematicians invented objects that contain geometrical/physical information about a manifold but are coordinate independent. Differential forms are examples of this type.

Finally a manifold can have features that remain unchanged even when the manifold is deformed preserving continuity. For example a doughnut cannot be changed into a sphere without tearing but it can be deformed to look like a cup with one handle. The features of a manifold that are preserved under continuous deformations are called topological invariants. Homotopy, Homology and Cohomology are groups that are topological invariants.

Thus the reason the topics listed are worth studying is that they make it possible to carry out calculus calculations on manifold, to present calculations in a coordinate independent way and highlight global features of a manifold that have important physical consequences.

Finally in order to study physical phenomena on a given a manifold always requires the introduction of another space. Suppose, for example, we want to study wind flows on the earth. Then the surface of the earth is our manifold to be described using patches but we also need to introduce a second space to describe the wind speed and direction. This space is three dimensional Euclidean space. Both these spaces are required in order to describe and model the physics system. The two spaces, the collection of two dimensional patches describing the earth, and the three dimensional Euclidean space describing the wind velocity, have to be glued together in a smooth way.

The construction described has been studied in great generality by mathematicians under the name: fibre bundle. We will use this idea to understand the topological insulator. This completes our overview of the mathematical topics listed. We now proceed to explain the key concepts that we need. Our aim is to quickly define terms and introduce useful calculational tools. We start with manifolds.

3.1   Manifolds

A manifold M is described by overlapping patches $\{U_\alpha\}$, called charts, that completely cover it and each chart has an invertible map $\{\Psi_\alpha\}$ from a portion of the manifold to Euclidean space. If a finite number of overlapping patches (charts) are enough to describe a manifold then it is said to be compact if an infinite number of patches are needed the manifold is said to be non compact.

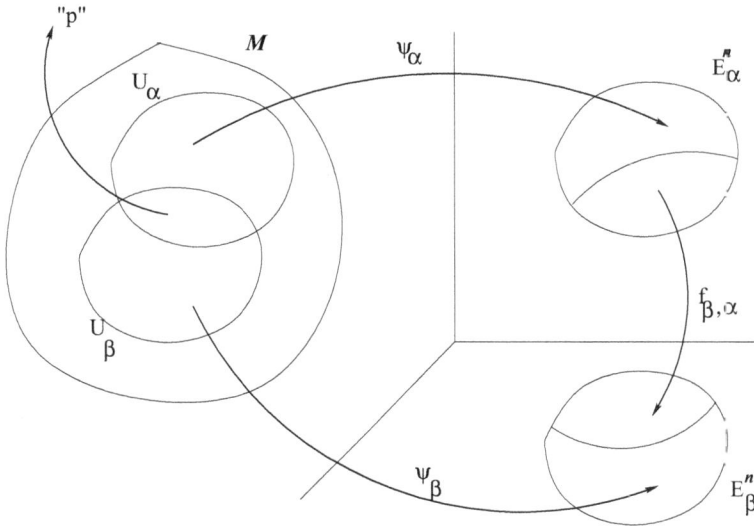

Fig. 3.1   Charts from M to Euclidean space.

Euclidean space is a manifold but it is not compact. However it can be covered by a countable infinite number of charts. Local charts can be defined in terms of spherical balls centred at points in Euclidean space. If the location of these centres is taken to be a collection of rational numbers then the number of balls will be countably infinite. The key point is that the rationals are dense in the reals which means that any real number is the limit of a sequence of rational numbers.

To clarify ideas let us consider a simple manifold, namely a circle S^1. There are many different ways of describing S^i. Algebraically we can think of it is points (x, y) such that $x^2 + y^2 = 1$ or as points on the real line where points x and $x + L$ are identified or as the space defined by the set of equations: $x^2 + y^2 + z^2 = 1$ and $z = 0$. Each one of these descriptions has an extension that can be useful for studying more general spaces. The method of defining S^1 where points x and $x + L$ are identified (i.e. glued together) is called a quotient construction.

We can introduce more structure to the manifold. We can represent S^1 as a differential manifold. This means that the gluing maps introduced to describe S^1 are chosen be differential not just continuous maps. Once this is done then we can introduce the operations of calculus on each patch and then use the smoothness property of the gluing maps to extended the operations of calculus to the entire manifold.

The first step is to introduce charts that cover S^1 and smooth gluing maps. Maps from these charts which give coordinates and the collection of charts provides a coordinate system for describing S^1. Then functions depending on coordinates can be introduced which can be differentiated and we have calculus on S^1. The technical details of this scheme will not be given. Our aim is to explain the ideas involved. In applications the precise form of coordinates is rarely required.

To describe S^1 a minimum of two coordinate charts are required. There are many ways of introducing charts. Let us describe one way. We take S^1 to be the set of points on **R** defined by $S^1 : \{(y, x) | x^2 + y^2 = 1\}$. A chart maps the open set U_1, defined to be the circle minus the point $\theta = 0$, where θ gives points on the circle and ranges between 0 and 2π to the Euclidean line (the x axis via a stereographic projection). We draw a straight line from the point $\theta = 0$ to the line $y = 0$, and get a point on the x axis, namely, the point E_p. Explicitly we have:

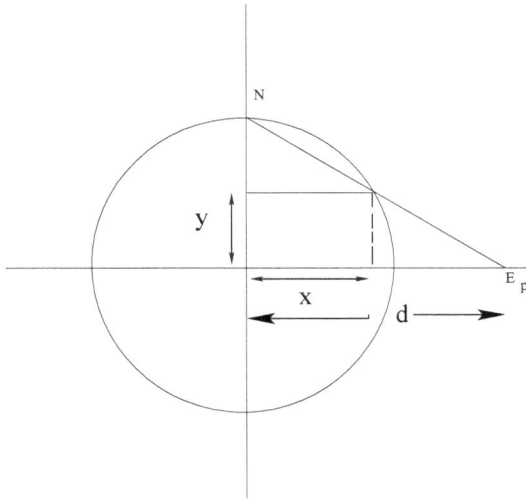

Fig. 3.2   Defining a stereographic coordinate chart on S^1.

$$\frac{x}{d} = \frac{1-y}{1},$$

$$d = \frac{x}{1-y}$$

The resulting coordinates cover the entire circle except the point $\theta = 0$. Similarly one can think of the chart which covers the circle except the point $\theta = 2\pi$. These two charts thus cover all points of the circle. The simple example of S^1 was given in order to show how charts can be introduced. There are many ways of introducing them. Once introduced they give a local description of a manifold in terms of Euclidean space and calculus can be introduced on the manifold.

3.1.1   *Differential Forms*

Differential forms and their dual, the vector field, are natural global objects that appear when the operations of calculus are introduced on a manifold. Functions are the simplest example of a differential forms. They are called differential zero forms. In general differential forms are anti-symmetric tensor fields that appear in physics and geometry. For example the electromagnetic field $F_{\mu\nu}$ and the curvature tensor $R_{\mu\nu}$ are a differential two forms, while vector fields appear as generators of change and symmetry.

Both objects can be defined in a coordinate independent way on a manifold. However they can also be introduced using a local Euclidean coordinates description. The local description is useful for calculations while the coordinate free description is useful for spotting general features of a problem. We start with a local description for a p-form, $\omega^{(p)}$ which is:

$$\omega^{(p)} = \sum \omega_{i_1, i_2, \cdots, i_p}(x) dx^{i_1} \wedge dx^{i_2} \wedge \cdots \wedge dx^{i_p},$$

$\omega^{(p)}$ is an object on the manifold, the expression written down gives a representation of this object using the coordinate building block differentials $dx^{i_1}, dx^{i_2}, \ldots$. These differentials satisfy the basic antisymmetric rule of $\wedge$ multiplication, $dx \wedge dy = -dy \wedge dx$. Note that in the summation each i_k is from $1, \cdots, n$. This leads to many identical terms in the expression above, since any permutation of the labels $i_1, \cdots, i_p$ give the same contribution. In view of the fact that an odd permutation will introduce a negative term in $\omega^{(p)}$, which will be compensated by a negative term from $dx^{i_1} \wedge \cdots dx^{i_p}$. Thus, there are $p!$ equal contributions to $\omega^{(p)}$. Very often this feature is recognized by introducing a $\frac{1}{p!}$ term. Thus we can write

$$\omega^{(p)} = \sum_{i_1 < \cdots < i_p} \omega_{i_1, \cdots, i_p}(x) dx^{i_1} \wedge \cdots \wedge dx^{i_p},$$

$$= \frac{1}{p!} \sum_{i_1, \cdots, i_p = 1}^{n} \omega_{i_1, \cdots, i_p}(x) dx^{i_1} \wedge \cdots \wedge dx^{i_p}$$

p-forms can be added and multiplied by scalars to generate new p-forms. They thus form a vector space. If the dimension of the manifold is n then differential one forms are elements of a n dimensional vector space with basis vectors $dx_i, i = 1, \ldots n$ while differential p-forms belong to a vector space of dimension ^nC_p with basis vectors constructed by taking the wedge product of p distinct one forms taken from $dx_1, dx_2, \ldots, dx_n$. Differential p and a q-form are also elements of what is called an exterior algebra as they can be multiplied. (wedge product) to generate a $(p + q)$-form. Explicitly

$$\omega^{(p)} \wedge \mu^{(q)} \rightarrow \Lambda^{(p+q)},$$

$$\omega^{(p)} \wedge \mu^{(q)} = (-1)^{pq} \mu^{(q)} \wedge \omega^{(p)} \quad : \text{non} - \text{commutative}$$

$$\mu_1^{q_1} \wedge (\mu_2^{q_2} \wedge \mu_3^{q_3}) = (\mu_1^{q_1} \wedge \mu_2^{q_2}) \wedge \mu_3^{q_3} \quad : \text{associative}.$$

Note : The wedge product is not commutative but it is associativity.

The coefficients $\omega_{i_1, i_2, \cdots, i_p}(x^i(p))$ in the local description of a p-form can be given a coordinate independent interpretation. We do not go into this detail.

3.1.1.1 *Examples*

In dimension two we have the following forms:

- Zero form: $f_0(x)$: function.
- One form: $f_1(x)dx^1 + f_2(x)dx^2$, and
- Two form: $f_3(x)dx^1 \wedge dx^2$.

In three dimensions we can have

- Zero form: f_0,
- One form: $f_1 dx^1 + f_2 dx^2 + f_3 dx^3$,
- Two form: $g_1 dx^1 \wedge dx^2 + g_2 dx^3 \wedge dx^1 + g_3 dx^2 \wedge dx^3$,
- Three form: $h\ dx^1 \wedge dx^2 \wedge dx^3$.

Now, let us return to the discussion of d operator. In two dimensions we can get

$$d(f_1 dx^1 + f_2 dx^2) = \frac{\partial f_1}{\partial x^2} dx^2 \wedge dx^1 + \frac{\partial f_2}{\partial x^1} dx^1 \wedge dx^2,$$

$$= -\left(\frac{\partial f_1}{\partial x^2} - \frac{\partial f_2}{\partial x^1}\right) dx^1 \wedge dx^2,$$

$$= \left(\frac{\partial f_2}{\partial x^1} - \frac{\partial f_1}{\partial x^2}\right) dx^1 \wedge dx^2.$$

Which is basically the "curl" (area).

Next consider the two form in dim 3:

$$d(g_1 dx^1 \wedge dx^2 + g_2 dx^3 \wedge dx^1 + g_3 dx^2 \wedge dx^3)$$

$$= \frac{\partial g_1}{\partial x^3} dx^3 \wedge dx^1 \wedge dx^2 + \frac{\partial g_2}{\partial x^2} dx^2 \wedge dx^3 \wedge dx^1 + \frac{\partial g_3}{\partial x^1} dx^1 \wedge dx^2 \wedge dx^3,$$

$$= \left(\frac{\partial g_3}{\partial x^1} + \frac{\partial g_2}{\partial x^2} + \frac{\partial g_1}{\partial x^3}\right) dx^1 \wedge dx^2 \wedge dx^3 \Rightarrow \text{``Divergence'' operator}.$$

3.1.2 *Vector Fields*

Vector fields and differential forms are dual objects. Differential forms describe antisymmetric tensor fields in a coordinate independent way while vector fields are operators that change coordinates and hence change functions and forms.

We explain vector fields by looking at an example. Suppose there is a moving point p on a manifold M of dimension n. We want to describe this motion. This can be done by using a local coordinate description of $p(t)$ as $x_i(p(t)), i = 1, 2..n$. As the point $p(t)$ moves it generates a curve on

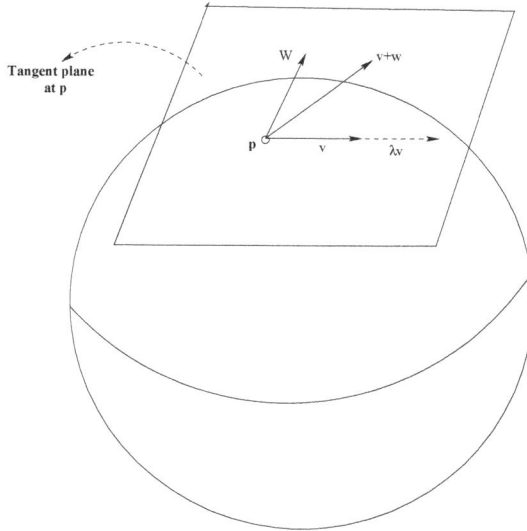

Fig. 3.3   Tangent plane at a point p to S^2.

the manifold M. Its velocity at time t is the tangent to this curve at the coordinate point $x_i(p(t))$, which is

$$\frac{dx_i(p(t))}{dt}, \ \ i = 1, 2, ..., n$$

Using this idea we consider the way a function $f(p)$ of $p(t)$ changes. We have

$$\frac{df}{dt} = \sum_i \frac{dx_i(p(t))}{dt} \frac{\partial f}{\partial x_i}$$

The next step is write this result as

$$\frac{df}{dt} = \mathbf{X}f$$

$$\mathbf{X} = \sum_i v_i \frac{\partial}{\partial x_i}$$

$$v_i = \frac{dx_i}{dt}$$

Thus we have introduced an operator $\mathbf{X}$ which changes a function. This operator is the vector field. Recall that two vector spaces are dual to one another if an element of one acting on an element of the other gives a scalar quantity. It was pointed out earlier that since differential forms of the same

order can be added and multiplied by a scalar to give a differential form
they are elements of a vector space. Similarly vector fields of the same
dimension form a vector space. These two vector spaces, when they are of
the same dimension, are a dual to one another which can be be displayed
as follows. We write

$$df = \frac{\partial f}{\partial x_i}$$
$$< df, \mathbf{X} > = \mathbf{X}f$$

In particular for $f = x_j$ and $\mathbf{X} = \frac{\partial}{\partial x_i}$ we have

$$< df, \mathbf{X} >=< dx_j, \frac{\partial}{\partial x_i} >= \frac{\partial x_j}{\partial x_i} = \delta_{i,j}$$

This result makes the dual link between forms and vector fields clear by
showing how they can be paired to give a number, $\delta_{i,j}$ which is equal to
one when $i = j$ and is zero otherwise. Very often we will refer to the vector
fields $\frac{\partial}{\partial x_i}$ as elements of the tangent space of a manifold and refer to dx_i as
elements of the dual cotangent space when $i = 1, ...n$ where the dimension
of the manifold is n.

We next introduce the fundamental notion of distance on a manifold.

3.1.3 Metric Tensor

The idea of length and distance is encoded in the metric tensor which is a
symmetric tensor field $g_{ij}(x)$. This tensor field provides a local description
of the square of the length of a line element which is:

$$g = ds^2 = \sum_{i,j} g_{ij}(x)dx^i dx^j.$$

A manifold on which it is possible to introduce such a metric tensor g is
called a Riemannian manifold and $g_{ij}(x)$ is the Riemannian metric of the
manifold. A general theorem states that a Riemannian metric can be given
to any paracompact manifold. Abstractly, the Riemannian line element
built out of two basis elements dx^i, dx^j of the cotangent space is a tensor
of type $(0, 2)$. A tensor of type (p, q) has q vector indices and p one form or
cotangent space indices. A coordinate independent description for g_{ij} can
be given by evaluating g on two vectors. Evaluation means

$$g\left(\frac{\partial}{\partial x^i}, \frac{\partial}{\partial x^j}\right) = \sum_{i,j} g_{ij} \left\langle dx^i dx^j, \frac{\partial}{\partial x^i} \otimes \frac{\partial}{\partial x^j} \right\rangle$$
$$= g_{ij}.$$

As in the case of p-forms the metric tensor g is built out of elements of the cotangent space. It is assumed to be globally defined on the manifold. The expression given for g represents it in a local coordinate chart.

We next consider the operations of calculus on forms. We start with the Lie derivative.

3.2   Operations on Forms

3.2.1   *Lie Derivative*

The Lie derivative determines the way a tensor fields defined on the manifold changes due to transformations generated by a vector. However a vector field acting on a form changes its order and hence is not directly useful for studying the way the form changes. We thus need an operator that acts on a form and gives back a form. If the form has a symmetry this operator should not change the form. The Lie derivative is such an operator.

To motivate how this search for symmetry and change point of view works let us consider a manifold with a translation invariant function $f(x)$ in the local coordinate direction x. This means that the function $f(x)$ has the property that $f(x) = f(x + th)$. But for a smooth function we can write $f(x + th) = e^{th\frac{d}{dx}f(x)}$, which for an infinitesimal t gives $f(x + th) = f(x) + th\frac{d}{dx}f(x)$. We recognize $h\frac{d}{dx}$ as a vector field X, which generates the symmetry if $Xf = 0$. Furthermore it is also clear that e^{tX} is an element of an Abelian group. The group elements being labelled by t with multiplication rule $e^{tX} \cdot e^{sX} = e^{t+s}X$. The identity element corresponds to $t = 0$, while the inverse element of e^{tX} is e^{-tX}. Thus translation invariance of $f(x)$ implies the existence of an Abelian group. We would like to extend this idea of symmetry and change to differential forms.

Before we do that we point out a special geometric feature for transformations generated by a vector field which has one parameter. For each value of this parameter we can suppose the vector field prescribes tangent direction on the manifold M. A theorem of mathematics then tells us that there is a curve on the manifold which has these prescribed tangents. However if we introduce two parameter vector fields on M there is, in general, no two dimensional surface on M with these prescribed tangent vectors. Additional constraints known as integrability conditions need to be satisfied.

Establishing the existence of a curve with prescribed tangent vectors is straightforward. We sketch the argument.

The change in coordinate functions of a point on a manifold with respect to the group parameter t for a general vector field X is

$$\frac{dx^i}{dt} = f^i(x^1, \cdots, x^n), \quad i = 1, \cdots, n$$

when

$$X = \sum_i f^i \frac{\partial}{\partial x^i}$$

This follows from the formula

$$\lim_{t \to 0} \frac{1}{t}(e^{tX} - 1)f(x) = Xf,$$

applied to the coordinate function x^i.

These are set of n-ordinary differential equations. The existence theorem for differential equations guarantees a local solution for $x^i(t)$ under suitable conditions on f^i. The solutions thus represent a curve on the manifold with prescribed tangent vector.

We now introduce the Lie derivative associated with a given vector fields and determine the way it act on functions and then use this result to determine the way it act on forms and on the metric.

3.2.1.1 *Action of Lie derivative on a function*

We write the Lie derivative $\mathcal{L}_X$ acting on a function f as

$$\mathcal{L}_X f = Xf,$$

where,

$$X = \sum_{i=1}^n v^i(x) \frac{\partial}{\partial x^i}.$$

This follows from

$$\mathcal{L}_X f = \lim_{t \to 0} \frac{1}{t}(e^{tX} - 1)f(x(p))$$

3.2.1.2 *Action of Lie derivative on a vector field*

We next determine the action of $\mathcal{L}_X$ on a vector field Y. We suppose,

$$\mathcal{L}_X(Yf) = (\mathcal{L}_X Y)f + Y(\mathcal{L}_X f),$$
$$\text{or,} \quad (\mathcal{L}_X Y)f = \mathcal{L}_X(Yf) - Y(\mathcal{L}_X f),$$

where Yf is a function, so $\mathcal{L}_X(Yf) = XYf$, similarly $Y(\mathcal{L}_Xf) = YXf$. Thus,

$$(\mathcal{L}_XY)f = XYf - YXf,$$
$$= [X,Y]f,$$

or,

$$\mathcal{L}_XY = [X,Y].$$

In doing this calculation we have assumed that the Lie derivative on the tensor product has the property

$$\mathcal{L}_X(T \otimes S) = (\mathcal{L}_XT) \otimes S + T \otimes (\mathcal{L}_XS).$$

We will establish the result shortly.

3.2.1.3   *Action of Lie derivative on the metric tensor*

We determine $\mathcal{L}_Zg(X,Y)$. We remember that $g(X,Y)$ is a function. It is basically the metric component g_{ij}.

$$\mathcal{L}_Z(g(X,Y)) = \mathcal{L}_Z(g, X \otimes Y).$$

The LHS of Eqn.

$$\mathcal{L}_Z(g(X,Y)) = Zg(X,Y),$$

while the RHS is

$$\mathcal{L}_Z(g, X \otimes Y) = (\mathcal{L}_Zg, X \otimes Y) + (g, \mathcal{L}_ZX \otimes Y) + (g, X \otimes \mathcal{L}_ZY),$$

Now, from Eqn and using $\mathcal{L}_ZY = [Z,Y]$, we get,

$$Zg(X,Y) = (\mathcal{L}_Zg, X \otimes Y) + (g, [Z,X] \otimes Y) + (g, X \otimes [Z,Y]),$$

or, $(\mathcal{L}_Zg, X \otimes Y) = Zg(X,Y) - (g, [Z,X] \otimes Y) - (g, X \otimes [Z,Y]).$

Now, let

$$X = \frac{\partial}{\partial x^i}, \quad Y = \frac{\partial}{\partial x^j}, \quad Z = \sum_{a=1}^{n} v^a(x)\frac{\partial}{\partial x^a}.$$

If $\mathcal{L}_Z$ is the isometry (i.e. it does not change the metric), then

$$(\mathcal{L}_Zg)(X,Y) = 0$$

the above equation is called the Killing Equation. It determines $v^a(x)$ for a given $g_{ij}(x)$. Hence Killing equation becomes

$$0 = Zg(X,Y) - (g, [Z,X] \otimes Y) - (g, X \otimes [Z,Y]),$$
$$= \sum_{a=1}^{n} v^a(x)\frac{\partial g_{ij}(x)}{\partial x^a} - g([Z,X],Y) - g(X,[Z,Y])$$

So, we have to evaluate 2nd and 3rd term of the right hand side of the killing equation. For that, we will use the following lemma:

$$[AB, C] = [A, C]B + A[B, C].$$

Hence,

$$[Z, X] = \sum_{a=1}^{n} \left[v^a(x) \frac{\partial}{\partial x^a}, \frac{\partial}{\partial x^i} \right],$$

$$= \sum_{a=1}^{n} \left\{ \left[v^a(x), \frac{\partial}{\partial x^i} \right] \frac{\partial}{\partial x^a} + v^a(x) \underbrace{\left[\frac{\partial}{\partial x^a}, \frac{\partial}{\partial x^i} \right]}_{=0} \right\}.$$

The first term of (3.1) can be written as

$$\left[v^a(x), \frac{\partial}{\partial x^i} \right] f = \left(v^a(x) \frac{\partial}{\partial x^i} - \frac{\partial}{\partial x^i} v^a(x) \right) f,$$

$$= v^a(x) \frac{\partial f}{\partial x^i} - \frac{\partial}{\partial x^i} (v^a(x) f),$$

$$= - \left(\frac{\partial v^a(x)}{\partial x^i} \right) f$$

So,

$$\left[v^a(x), \frac{\partial}{\partial x^i} \right] = - \left(\frac{\partial v^a(x)}{\partial x^i} \right).$$

Hence, from Eq. (3.1)

$$[Z, X] = - \sum_{a=1}^{n} \left(\frac{\partial v^a(x)}{\partial x^i} \right) \frac{\partial}{\partial x^a}.$$

$$\text{and, } [Z, Y] = - \sum_{a=1}^{n} \left(\frac{\partial v^a(x)}{\partial x^j} \right) \frac{\partial}{\partial x^a}.$$

So, finally we get

$$0 = \sum_{a=1}^{n} v^a(x) \frac{\partial g_{ij}(x)}{\partial x^a} + \sum_{a=1}^{n} g \left(\frac{\partial v^a(x)}{\partial x^i} \frac{\partial}{\partial x^a}, \frac{\partial}{\partial x^j} \right)$$

$$+ \sum_{a=1}^{n} g \left(\frac{\partial}{\partial x^i}, \frac{\partial v^a(x)}{\partial x^j} \frac{\partial}{\partial x^a} \right)$$

$$\text{or, } = \sum_{a=1}^{n} \left[v^a(x) \frac{\partial g_{ij}(x)}{\partial x^a} + \frac{\partial v^a}{\partial x^i} g_{aj}(x) + \frac{\partial v^a}{\partial x^j} g_{ia}(x) \right]$$

3.2.1.4 *Lie derivative of one form*

We proceed to determine the action of the Lie derivative on a p form.

$$\mathcal{L}_Z(\omega, X) = (\mathcal{L}_Z\omega, X) + (\omega, \mathcal{L}_Z X),$$
$$\text{now, } (\mathcal{L}_Z\omega, X) = \mathcal{L}_Z(\omega X) - (\omega, \mathcal{L}_Z X),$$

We recall that locally for a one form,

$$(\omega, X) = \sum_{a=1}^{n} \left(\omega_a dx^a, \frac{\partial}{\partial x^i} \right) = \omega_i(x),$$

$$\text{hence, } \mathcal{L}_Z(\omega X) = Z\omega_i(x)$$

Hence, we get

$$(\mathcal{L}_Z\omega, X) = Z\omega_i(x) - (\omega, [Z, X]),$$
$$= Z\omega_i(x) + \sum_{a=1}^{n} \left(\omega_j dx^j, \frac{\partial v^a}{\partial x^i} \frac{\partial}{\partial x^a} \right),$$
$$= \sum_{a=1}^{n} \left(v^a \frac{\partial \omega_i(x)}{\partial x^a} + \omega_a(x) \frac{\partial v^a}{\partial x^i} \right)$$

For a general p-form the corresponding result is

$$(\mathcal{L}_Z\omega^{(p)})(X_1, \cdots, X_p) = Z(\omega^{(p)}(X_1, \cdots, X_p))$$
$$- \sum_{k=1}^{p} \omega^{(p)}(X_1, \cdots, [Z, X_k], \cdots, X_p).$$

A simple consequence of this result is

$$\mathcal{L}_X(T \otimes S) = (\mathcal{L}_X T) \otimes S + T \otimes (\mathcal{L}_X S).$$

We note that

$$(\mathcal{L}_Z(T^{(p)} \otimes S^{(q)}))(X_1, \cdots, X_{p+q}) = Z((T^{(p)} \otimes S^{(q)})(X_1, \cdots, X_{p+q}))$$
$$- \sum_{k=1}^{p+q} (T^{(p)} \otimes S^{(q)})(X_1, \cdots, [Z, X_k], \cdots, X_{p+q}).$$

from the result just stated. On the other hand by definition

$$(T^{(p)} \otimes S^{(q)})(X_1, \cdots, X_{p+q}) = T^{(p)}(X_1, \cdots, X_p) S^{(q)}(X_{p+1}, \cdots, X_{p+q}).$$

Hence

$$Z((T^{(p)} \otimes S^{(q)})(X_1, \cdots, X_{p+q})) = Z(T^{(p)}(X_1, \cdots, X_p)) S^{(q)}(X_{p+1}, \cdots, X_{p+q})$$
$$+ T^{(p)}(X_1, \cdots, X_p) Z(S^{(q)}(X_{p+1}, \cdots, X_{p+q})).$$

while the term with the commutator splits into two terms depending on whether $k < p$ or $k > p$, i.e.

$$= -\sum_{k=1}^{p} T^{(p)}(X_1, \cdots, [Z, X_k], \cdots, X_p) S^{(q)}(X_{p+1}, \cdots, X_{p+q})$$

$$-T^{(p)}(X_1, \cdots, X_p)(\sum_{k=p+1}^{p+q} S^{(q)}(X_{p+1}, \cdots, [Z, X_k], \cdots, X_{p+q})),$$

which establishes the result.

A consequence of this result is

$$\mathcal{L}_Z(\omega^{(p)} \wedge v^{(q)}) = \mathcal{L}_Z\omega^{(p)} \wedge v^{(q)} + \omega^{(p)} \wedge \mathcal{L}_Z v^{(q)}.$$

We next introduce three important operators.

3.3   Three New Operations: d, i_X and $\star$

There are three natural operations that act on forms and change their order. These operators can be used to construct physically important differential operators such as the Laplacian and the Lie derivative. The Laplacian shows up in many areas of physics and geometry and the Lie derivative can be used, as we explained, to probe symmetry. The three operators that change the order of a form are,

- The exterior derivative : $d : d\omega^{(p)} \to \omega^{(p+1)}$
- the interior product : $i_X : i_X\omega^{(p)} \to \omega^{(p-1)}$ and
- the Hodge star : $\star : \star\omega^{(p)} \to \omega^{(n-p)}$, where dim $\mathcal{M} = n$.

Two special operators can be constructed as products of the operators listed which do not change the order of a form. They are the Laplacian $\nabla_p = d^\dagger d + dd^\dagger$ where $d^\dagger = \star d\star$ and the Lie derivative which we have already discussed. However it was not shown that the Lie derivate can be written as $\mathcal{L} = i_X.d + d.i_X$. This will be done later.

An important property of differential forms is that they are metric independent. The operator d, called the exterior derivative, increases the order of a form by one is also metric independent. However the Hodge star operator which relates a p-form to its "dual": an $(n-p)$-form where n is the dimension of the manifold does depend on the metric of the manifold. The metric independent nature of forms makes them suitable to study topological features of a manifold. We will give examples of how this is done as we proceed.

3.3.1   *The Exterior Derivative (d)*

The exterior derivative maps a p-form to a $(p+1)$-form. Locally a p-form is represented in terms of the wedge product of p-terms. The action of the d-operator increases the number of such terms by one in the following way

$$d\omega^{(p)} = \sum_{i_1,\cdots,i_p,j} \frac{\partial\omega^{(p)}_{i_1,\cdots,i_p}(x)}{\partial x^j} dx^j \wedge dx^{i_1} \wedge \cdots \wedge dx^{i_p}.$$

An immediate important consequence of this representation is the result

$$d^2 = 0.$$

To see this, we simply use the definition of the d-operator. Although simple this result has important hidden features which we now discuss.

The result suggests the question: if $d\omega^{(p)} = 0$, does it mean that we must have $\omega^{(p)} = d\eta^{(p-1)}$ for some $(p-1)$-form η? Forms for which $d\omega^{(p)} = 0$ are called closed. While closed forms which can be written as $\omega^{(p)} = d\eta^{(p-1)}$ are called exact. We will see that locally all closed forms are exact, but this need not be the case globally. The local result is known as the Poincaré lemma while the possibility of the result not holding due to global considerations is a subject matter of de Rham cohomology where closed forms modulo exact forms are shown to lead to groups that are topological invariants of the manifold. They are cohomology groups of the manifold. Let us look at an example.

3.3.2   *A Brief Discussion on de-Rham Cohomology*

We now show that Poincaré's lemma need not hold globally. To do this we introduce two spaces, namely $\mathcal{Z}^p$ and $\mathcal{B}^p$ which consists of all closed and exact p-forms respectively on $\mathcal{M}$ defined as stated as follows

$\mathcal{Z}^p : \{\omega^{(p)} \mid d\omega^{(p)} = 0\} \Rightarrow$ this is called space of all closed p − forms,

$\mathcal{B}^p : \{\omega^{(p)} \mid \omega^{(p)} = d\eta^{(p-1)}\} \Rightarrow$ this is called space of all exact p − form.

Since an exact form is always closed, we can write

$$\mathcal{B}^p \subset \mathcal{Z}^p$$

One can define the quotient space as

$$\mathcal{H}^{(p)}(\mathcal{M}, \mathbf{R}) = \mathcal{Z}^p(\mathcal{M}, \mathbf{R})/\mathcal{B}^p(\mathcal{M}, \mathbf{R})$$
$$= p \text{ th de Rham cohomology group}$$

The group property can be seen easily if we suppose that the number of closed p-forms on a manifold is finite. One can then choose a basis of these forms. The coefficients of these basis p-forms can be chosen as arbitrary real numbers, i.e. they are elements of $\mathbf{R}$. Elements of $\mathbf{R}$ form an Abelian group under addition with zero as its identity element. Thus one can associate with $\mathcal{Z}^p$ and $\mathcal{B}^p$, Abelian groups which reflect the number of elements that are closed or exact. The quotient of these groups is the de Rham cohomology group. If Poincaré's lemma was globally valid, the spaces $\mathcal{Z}^p$ and $\mathcal{B}^p$ would be equal and the cohomology group would be trivial, i.e. it consists of a single element corresponding to the identity element of $\mathbf{R}$, which is zero.

We now give an example where this is not the case. This is the case of the one dimensional manifold S^1. For this manifold zero forms and one forms are only possible. An arbitrary zero form will be described by a function parametrized by an angle θ which is periodic, while a one form can be written as $\omega^{(1)} = f_1(\theta)d\theta$, again with $f_1(\theta)$ being periodic in θ. We now examine the space $\mathcal{Z}^1$ and $\mathcal{B}^1$.

$$\mathcal{Z}^1 = \{\omega^{(1)} \mid d\omega^{(1)} = 0\},$$
$$\mathcal{B}^1 = \{\omega^{(1)} \mid \omega^{(1)} = d\omega^{(0)}\}.$$

Let, $\omega^{(1)} = f_1(\theta)d\theta$, where $0 \le \theta \le 2\pi$. We write $f_1(\theta)$ as

$$f_1(\theta) = \sum_{n=-\infty}^{\infty} c_1^n e^{in\theta}.$$

Then $d\omega^{(1)} = 0$ trivially. We note from the above equations that

$$\omega^{(0)} = \sum_{n=-\infty}^{\infty} c_0^n e^{in\theta},$$

$$\text{or,} \quad d\omega^{(0)} = \sum_{n=-\infty}^{\infty} (in)c_0^n e^{in\theta} d\theta.$$

Since we want to find out all the exact forms, we need $\omega^{(1)} = d\omega^{(0)}$. Hence comparing coefficients, we get

$$c_1^n = c_0^n(in), \text{ for } n \ne 0,$$

which means that c_0^n, which is a real number can not be eliminated by a term coming from $d\omega^{(0)}$. Thus we can write

$$\mathcal{H}^{(1)}(S^1, \mathbf{R}) = \mathcal{Z}^1(S^1, \mathbf{R})/\mathcal{B}^1(S^1, \mathbf{R})$$
$$= c_n^0.$$

Thus we write

$$\mathcal{H}^1(S^1, \mathbf{R}) = \mathbf{R}.$$

Here $\mathcal{Z}^1 \neq \mathcal{B}^1$, i.e. Poincaré lemma does not extend globally in this case.

3.3.2.1   *Betti numbers*

We next introduce the important idea of a Betti number . The cohomology group for S^1 we found was a group generated by just one independent real number, c_n^0. For this reason it is said to have rank one. The rank of a cohomology group, an integer number, is called the Betti number. Thus for the circle S^1 the the cohomology group $\mathcal{H}^1(S^1)$ has of rank one and has Betti number $b_1(S^1) = 1$. Intuitively the Betti number counts the number of holes present in the manifold.

   For a two torus T^2, the surface of a doughnut, the manifold is $T^2 = S^1 \times S^1$. The de Rham cohomology groups of T^2 can be calculated by constructing the number of independent zero, one and two dimensional forms that are closed but not exact following the procedure described. The cohomology groups are

$$\mathcal{H}^0(T^2) = \mathbf{R}$$
$$\mathcal{H}^1(T^2) = \mathbf{R} + \mathbf{R}$$
$$\mathcal{H}^2(T^2) = \mathbf{R}$$

The associated Betti numbers of T^2 are: $b_0 = 1, b_1 = 2, b_2 = 1$. There are tables listing cohomology groups and Betti numbers for standard manifolds, such as spheres and tori of arbitrary dimension. We list Betti numbers for the S^n, the n-sphere $b_n = 1, b_0 = 1$ and all other Betti numbers $b_i = 0$

3.3.2.2   *The Kunneth Formula*

There is a useful formula for calculating Betti numbers for a manifold $M = A \times B$ from knowledge of the Betti numbers for A, B known as the Kunneth formula. It is

$$b_p(M) = \sum b_k(A) b_{p-k}(B)$$

This can be used to calculate the Betti numbers of, for instance, $T^2 = S^1 \times S^1$ from a knowledge of the Betti numbers of S^1.

3.3.2.3 *Action of d-operator on wedge products*

We next state Liebnitz rule for forms. It tells us the way the d operator acts on the wedge product of two forms. The result can be established using the local expressions for forms given. We have

$$d(\omega^{(p)} \wedge \mu^{(q)}) = (d\omega^{(p)}) \wedge \mu^{(q)} + (-1)^p \omega^{(p)} \wedge (d\mu^{(q)})$$

Differential forms and volume

Differential forms are closely linked to areas and volumes but without a length scale present. We describe the way metric information can be introduced. The trick is to introduce a one form from the metric. Let us explain how this done. A natural definition for the infinitesimal volume element is to require that it be proportional to $\epsilon_{1...n} e^1 \wedge \cdots \wedge e^n$ if the differentials e^i represent orthogonal directions in the space, i.e. if the corresponding metric can be written as

$$ds^2 = (e^1)^2 + \cdots + (e^n)^2.$$

The constant epsilon tensor introduced reflects the orientation of the volume element. We would like to determine the infinitesimal volume element defined for the general line element $ds^2 = g_{\mu\nu}(x)dx^\mu dx^\nu$, which follows from this natural definition.

$$ds^2 = \sum_{\mu,\nu} g_{\mu\nu}(x)dx^\mu dx^\nu,$$

The trick is to write

$$ds^2 = \sum_a (e^a)^2,$$

by setting

$$e^a = \sum_\mu e_\mu{}^a dx^\mu,$$

so that,

$$g_{\mu\nu} = \sum_a e_{a\mu} e_\mu^a$$

where, $\mu, \nu = 1, \cdots, n$ and $a, b = 1, \cdots, n$. The volume element is given by

$$e^1 \wedge \cdots \wedge e^n = e_{\mu_1}^{\ 1} \cdots e_{\mu_n}^{\ n} dx^{\mu_1} \wedge \cdots \wedge dx^{\mu_n} \epsilon_{\mu_1 \cdots \mu_n},$$
$$= \det(e_\mu{}^a) dx^1 \wedge \cdots \wedge dx^n.$$

But,

$$\det(g) = \det e^2$$
$$\Rightarrow \det e = \sqrt{\det g}$$

Thus the required volume element corresponding to the metric $g_{\mu\nu}$ is

$$\sqrt{\det g} dx^1 \wedge \cdots \wedge dx^n.$$

3.3.2.4 *Interior product*

The i_X-operator, known as the interior product, acting on a p-form, was introduced as a map that changed the p-form to a $(p-1)$-form. i.e. i_X : $i_X\omega^{(p)} \to \omega^{(p-1)}$, where X is a vector field given by $X = \sum_a f^a \frac{\partial}{\partial x^a}$. Let us explain how this is done. The idea is to "contract" the components of the vector field X_i with one of the corresponding dx^i-factor present in the local representation of a p-form in an antisymmetric way. Thus given a local expression for a p-form:

$$\omega^{(p)} = \sum_{i_1,\cdots,i_p} \omega_{i_1,\cdots,i_p}(x)dx^{i_1} \wedge \cdots \wedge dx^{i_p}.$$

$i_X\omega^{(p)}$ is defined as

$$i_X\omega^{(p)} = \sum_{i_1,\cdots,i_p,a} f^a(x)\omega_{i_1,\cdots,a,\cdots,i_p}(x)(-1)^a dx^{i_1} \wedge \cdots \wedge \widehat{dx^a} \wedge \cdots \wedge dx^{i_p}.$$

As always, the "carret" $\widehat{}$ over dx^a implies that the entry is missing. An immediate consequence of this definition is

$$\boxed{i_X^2\omega^{(p)} = 0},$$

since

$$i_X^2\omega^{(p)}$$
$$= \sum_{i_1,\cdots,i_p,a,b} f^a(x)f^b(x)\omega_{i_1,\cdots,a,\cdots,b,\cdots,i_p}(-1)^a(-1)^b dx^{i_1} \wedge \cdots$$
$$\wedge \widehat{dx^a} \wedge \cdots \wedge \widehat{dx^b} \wedge \cdots \wedge dx^{i_p}.$$

As ω is anti-symmetric but $f^a(x)f^b(x)$ is symmetric, hence i_X^2 is zero.

3.3.2.5 *Action of i_X-operator on wedge products*

Again using the local expression for $\omega^{(p)}$ and $\mu^{(q)}$ we proceed to establish

$$i_X(\omega^{(p)} \wedge \mu^{(q)}) = i_X\omega^{(p)} \wedge \mu^{(q)} + (-1)^p \omega^{(p)} \wedge i_X\mu^{(q)}$$

To prove this we note

$$i_X(\omega^{(p)} \wedge \mu^{(q)})$$

$$= i_X\left[\sum_{i_1,\cdots,i_{p+q}} \omega_{i_1,\cdots,i_p}(x)dx^{i_1} \wedge \cdots \wedge dx^{i_p} \wedge \cdots \wedge dx^{i_{p+q}}\mu_{i_{p+1},\cdots,i_{p+q}}(x) \right]$$

$$= \sum_{i_1,\cdots,i_{p+q}} [(-1)^a f^a(x)\omega_{i_1,\cdots,a,\cdots,i_p}(x)\mu_{i_{p+1},\cdots,i_{p+q}}(x)dx^{i_1} \wedge \cdots \wedge \widehat{dx^a} \wedge$$

$$\cdots \wedge dx^{i_p} \wedge dx^{i_{p+1}} \wedge \cdots \wedge dx^{i_{p+q}} + (-1)^b \omega_{i_1,\cdots,i_p}(x)\mu_{i_{p+1},\cdots,b,\cdots,i_{p+q}}(x)$$

$$dx^{i_1} \wedge \cdots \wedge dx^{i_p} \wedge dx^{i_{p+1}} \wedge \cdots \wedge \widehat{dx^b} \wedge \cdots \wedge dx^{i_{p+q}}]$$

where $(-1)^b = (-1)^{p+c}$ where c is form i_{p+1} to i_{p+q}.

Hence

$$i_X\omega^{(p)} \wedge \mu^{(q)} + \sum_{i_1,\cdots,i_{p+q}} (-1)^p\omega_{i_1,\cdots,i_p}(x)\mu_{i_{p+1},\cdots,i_{p+q}}(x)(-1)^c$$

$$dx^{i_1} \wedge \cdots \wedge dx^{i_p} \wedge dx^{i_{p+1}} \wedge \cdots \wedge dx^{i_{p+q}},$$

$$= i_X\omega^{(p)} \wedge \mu^{(q)} + (-1)^p\omega^{(p)} \wedge i_X\mu^{(q)}.$$

Finally we note that there is a simple coordinate independent definition for the action of interior product operator i_X on a p-form $\omega^{(p)}$, namely

$$(i_X\omega^{(p)})(X_1,\cdots,X_{p-1}) = \omega^{(p)}(X,X_1,\cdots,X_{p-1})$$

3.3.2.6 *Hodge star operator*

The Hodge star operator depends on metric. It maps a p-form to an $(n-p)$-form where n is the dimension of the manifold. We recall that the space of p-forms in an n-dimensional manifold is a vector space $\Lambda^{(p)}$ of dimension ^nC_p. The space of $(n-p)$-forms for the same manifold is a vector space $\Lambda^{(n-p)}$ of dimension $^nC_{n-p}$. Thus

$$\dim \Lambda^{(n-p)} = {}^nC_{n-p} = \dim \Lambda^{(p)}.$$

This suggests that a mapping between these spaces is possible. The basic geometrical idea is that using $\omega^{(p)}$ and taking its wedge product with $\omega^{(n-p)}$ gives a top form, which is one dimensional and can be interpreted as being proportional to local volume element. Thus the star product can be implicitly defined in a coordinate independent way. Note the explicit dependence on the metric through the appearance of the volume term τ. We can write τ using the anti-symmetric properties of the wedge product as

$$\tau = \sum_{i_1,\cdots,i_n} \frac{1}{n!}\sqrt{|g|}\epsilon_{i_1\cdots i_n}dx^{i_1}\wedge\cdots\wedge dx^{i_n},$$

$$= \sum_{i_{p+1},\cdots i_n} \frac{1}{(n-p)!}\sqrt{|g|}\epsilon_{12\cdots p i_{p+1}\cdots i_n}dx^1\wedge\cdots\wedge dx^p\wedge dx^{i_{p+1}}\wedge\cdots\wedge dx^{i_n}.$$

Note, that the factorial terms that are there because of symmetry, for instance there are $n!$ identical contributions corresponding to the $n!$ orderings that are possible in $dx^{i_1}\wedge\cdots dx^{i_n}$. By choosing an ordering of the labels $i_1 < \cdots < i_n$ introduces a restriction on the summation of terms. There is no longer the $n!$ symmetry property. Hence for summations over variable $i_1,\ldots,i_n$ restricted by this ordering and there is no need to divide by $n!$. In this section we will use this property to carry out calculations without symmetry. At the final step we can always remove restrictions on the summations present and reintroduce an appropriate factorial.

$$\star^2 = (-1)^{n(n-p)}$$

We have seen that for the Hodge star operator

$$\star\omega^{(p)}\rightarrow\omega^{(n-p)}$$
$$\star\omega^{(n-p)}\rightarrow\omega^{(n-(n-p))}=\omega^{(p)}$$
$$\text{i.e. } \star\star\omega^{(p)}\rightarrow\omega^{(p)}.$$

Explicitly

$$\star(dx^{i_1}\wedge\ldots\wedge dx^{i_p}) = \frac{\sqrt{g}}{(n-p)!}\epsilon^{i_1,\ldots,i_p}_{i_{p+1},\ldots,i_n}dx^{i_{p+1}}\wedge\ldots\wedge dx^{i_n}.$$

So,

$$\star\star(dx^{i_1}\wedge\cdots\wedge dx^{i_p}) = \frac{\sqrt{g}}{(n-p)!}\epsilon^{i_1\cdots i_p}{}_{i_{p+1}\cdots i_n}\star(dx^{i_{p+1}}\wedge\cdots\wedge dx^{i_n}),$$

$$= \frac{\sqrt{g}}{(n-p)!}\epsilon^{i_1\cdots i_p}{}_{i_{p+1}\cdots i_n}\frac{\sqrt{g}}{p!}\epsilon^{i_{p+1}\cdots i_n}{}_{j_1\cdots j_p}$$
$$dx^{j_1}\wedge\cdots\wedge dx^{j_p}$$
$$= \frac{\sqrt{g}}{(n-p)!}\frac{\sqrt{g}}{p!}g^{i_1 k_1}\cdots g^{i_p k_p}\epsilon_{k_1\cdots k_p i_{p+1}\cdots i_n}$$
$$g^{i_{p+1}k_{p+1}}\cdots g^{i_n k_n}\epsilon_{k_{p+1}\cdots k_n j_1\cdots j_p}dx^{j_1}\wedge\cdots\wedge dx^{j_p}$$
$$= \frac{\sqrt{g}}{(n-p)!}\frac{\sqrt{g}}{p!}g^{i_1 k_1}\cdots g^{i_p k_p}g^{i_{p+1}k_{p+1}}\cdots g^{i_n k_n}$$
$$(-1)^{n(n-p)}\epsilon_{k_1\cdots k_p i_{p+1}\cdots i_n}\epsilon_{j_1\cdots j_p k_{p+1}\cdots k_n}$$
$$dx^{j_1}\wedge\cdots\cdots\wedge dx^{j_p}$$

Hence

$$\boxed{\star\star(dx^{i_1}\wedge\cdots\wedge dx^{i_p}) = (-1)^{n(n-p)}dx^{i_1}\wedge\cdots\cdots\wedge dx^{i_p}}.$$

3.3.2.7 *An example*

Consider 2-dimensional Euclidean space for which the metric is $g_{ij}=\delta_{ij}$ and $g^{ij}=\delta_{ij}$. Hence $\sqrt{|g|}=1$. Since $\epsilon_{12}=1$, we have the following actions of the Hodge star operator on the bases of the space

$$\star 1 = dx^1\wedge dx^2.$$

Similarly,

$$\star dx^1 = dx^2,$$

$$\star dx^2 = -dx^1,$$

and,

$$\star(dx^1\wedge dx^2) = 1.$$

3.3.3 *Natural Operators: Lie Derivative and Laplacian*

There are two natural operators which can be constructed from the operators d, i_X and $\star$, which map a p-form back to a p-form. We will show that these operates are nothing but the Lie derivative and the Laplacian operator. First we have the operator

$$i_X d + d i_X.$$

This operator depends linearly on the operator d. We will show that this operator is nothing but the Lie Derivative $\mathcal{L}_X$ (Cartan's formula). $\mathcal{L}_X = i_X d + d i_X$.

To do this, we simply show that

$$\mathcal{L}_X \omega^{(p)} = i_X d \omega^{(p)} + d i_X \omega^{(p)}.$$

From the definition of i_X, it follows that

$$(i_X d \omega^{(p)})(X_1, \cdots, X_p) = (d \omega^{(p)})(X, X_1, \cdots, X_p),$$

we then use the expression derived earlier for $d \omega^{(p)}(X_1, \cdots, X_p)$ to write

$$= X(\omega^{(p)}(X_1, \cdots, X_p)) + \sum_{j=1}^{p} (-1)^{j+2} \omega^{(p)}([X, X_j], \widehat{X}, \cdots, \widehat{X_j}, \cdots, X_p)$$

$$+ \sum_{i=1}^{p} (-1)^{i+2} X_i(\omega^{(p)}(X, X_1, \cdots, \widehat{X_i}, \cdots, X_p))$$

$$+ \sum_{i<j} (-1)^{i+j+2} \omega^{(p)}([X_i, X_j], X, \cdots, \widehat{X_i}, \cdots, \widehat{X_j}, \cdots, X_p).$$

The first two terms represent $\mathcal{L}_X$, defined earlier in a coordinate free way, while the last two terms combined represent $-(d i_X \omega^{(p)}(X_1, \cdots, X_p))$. The sign factors take into account the presence of the extra vector field X.

Secondly we have the operator

$$d(\star d \star) + (\star d \star) d.$$

This again maps $\omega^{(p)} \to \omega^{(p)}$. Explicitly we have the sequence

$$d \omega^{(p)} \to \omega^{(p+1)},$$
$$\star d \omega^{(p)} \to \omega^{n-(p+1)},$$
$$d \star d \omega^{(p)} \to \omega^{n-(p+1)+1},$$
$$\star d \star d \omega^{(p)} \to \omega^{n-[n-(p+1)+1]} = \omega^{(p)}.$$

similarly one can check that

$$d \star d \star \omega^{(p)} \to \omega^{(p)}$$

We will now show that this operator acting on zero forms i.e. functions is nothing but the Laplacian operator. The operator introduced can thus be interpreted as the generalization of the Laplacian operator appropriate for acting on forms.

We separately evaluate each term of the expression

$$(d \star d \star + \star d \star d)f,$$

The first term gives zero. Since we know that $\star f$ should give us a n-form (since dim $\mathcal{M} = n$), and hence $d \star f = 0$.

Now, let us look at the second term.

$$df = \frac{\partial f}{\partial x^\mu} dx^\mu,$$

$$\star df = \frac{\partial f}{\partial x^\mu} \star dx^\mu = \frac{\partial f}{\partial x^\mu} \sqrt{|g|} \epsilon^\mu{}_{i_2 \cdots i_n} dx^{i_2} \wedge \cdots \wedge dx^{i_n},$$

$$= \frac{\partial f}{\partial x^\mu} \sqrt{|g|} g^{\mu i_1} \epsilon_{i_1 i_2 \cdots i_n} dx^{i_2} \wedge \cdots \wedge dx^{i_n},$$

$$d \star df = \frac{\partial}{\partial x^\lambda} \left[\frac{\partial f}{\partial x^\mu} \sqrt{|g|} g^{\mu i_1} \right] \epsilon_{i_1 \cdots i_n} dx^\lambda \wedge dx^{i_2} \wedge \cdots \wedge dx^{i_n},$$

$$= \frac{\partial}{\partial x^{i_1}} \left[\frac{\partial f}{\partial x^\mu} \sqrt{|g|} g^{\mu i_1} \right] \epsilon_{i_1 \cdots i_n} dx^{i_1} \wedge dx^{i_2} \wedge \cdots \wedge dx^{i_n},$$

$$\star d \star df = \frac{\partial}{\partial x^{i_1}} \left[\frac{\partial f}{\partial x^\mu} \sqrt{|g|} g^{\mu i_1} \right] \epsilon_{i_1 \cdots i_n} \star (dx^{i_1} \wedge dx^{i_2} \wedge \cdots \wedge dx^{i_n}),$$

$$= \frac{\partial}{\partial x^{i_1}} \left[\frac{\partial f}{\partial x^\mu} \sqrt{|g|} g^{\mu i_1} \right] \epsilon_{i_1 \cdots i_n} g^{i_1 j_1} \cdots g^{i_n j_n} \epsilon_{j_1 \cdots j_n} \sqrt{|g|},$$

$$= \frac{\partial}{\partial x^{i_1}} \left[\frac{\partial f}{\partial x^\mu} \sqrt{|g|} g^{\mu i_1} \right] \underbrace{\epsilon_{i_1 \cdots i_n} \epsilon^{i_1 \cdots i_n}}_{\frac{1}{|g|}} \sqrt{|g|},$$

$$= \frac{1}{\sqrt{|g|}} \frac{\partial}{\partial x^{i_1}} \left[\frac{\partial f}{\partial x^\mu} \sqrt{|g|} g^{\mu i_1} \right]$$

We can denote $\star d \star$ by $d^\dagger$, then the Laplacian is

$$\nabla_p^2 \equiv \Delta_p = (d^\dagger d + dd^\dagger).$$

We have described the basic working tools needed to do calculations with forms. It has been suggested that geometrical and topological methods will play a more central role in condensed matter physics in the immediate future. It is for this reason that we have given calculational rules involving forms in some detail. The short account we have given summarises important ideas of differential geometry and topology should be helpful for following more advanced arguments. However for physics applications it is most important to have a clear intuitive understanding of key concepts of a mathematical topic and to have the ability to carry out relevant calculations. Complete mastery of the technical aspects of a proof are not as important.

3.4   Integration on a Manifold $\mathcal{M}$

We next turn to integration where we state Stoke's theorem. This theorem is a very important for integrating a form on a manifold with a boundary.

Stoke's Theorem states that the integral of a $(n-1)$-form $\omega^{(n-1)}$ over a $(n-1)$-dimensional boundary $\partial\mathcal{M}$ is equal to the integral of $d\omega^{(n-1)}$ over the n-dimensional manifold which has $\partial\mathcal{M}$ as its boundary.

$$\int_{\mathcal{M}} d\omega^{(n-1)} = \int_{\partial\mathcal{M}} \omega^{(n-1)}$$

We do not prove the result in detail but sketch how it his can be done. Two ideas are required. The first is to know how to represent a boundary in terms of Euclidean space and coordinates. The second is to find a way to reduce calculations to one involving just local coordinate charts. Once this is done integration on the manifold is reduced to the sum of integrals over local charts.

Let us examine the notion of a boundary in the setting of setting of Euclidean space. We can motivate the procedure we follow by considering a two dimensional example. In two dimensions, we can map small regions into a square which is labeled by Euclidean variables x, y, where $-l < x < l$ and $-l < y < l$. The boundary region can be represented geometrically by choosing, for example, the boundary curve to be the x-axis. Now, x will again range between $-l$ and l, but y is restricted to lie between 0 and l as it ends on the x axis. The generalization to higher dimensions (say n) replaces the square by a n-dimensional hypercube with Cartesian coordinates $(x^1, \cdots, x^n)$, all ranging between $-l$ and l, while the boundary is represented by coordinates $x^1, \cdots, x^n$ in a region defined by $-l < x^1, \cdots x^{n-1} < l$ and $0 < x^n < l$. By a suitable choice of l such a collection of elements from an Euclidean space can be used to describe the given manifold.

Thus locally a Euclidean space description of a boundary can be given. Integration over these local Euclidean spaces can be done. Our task is to find a way to add up these integrals over local charts to get the integral over the manifold, An extremely clever method for doing this was found. It is called a partition of unity subordinate to the cover used to describe the manifold. Let us explain what this means. The partition of unity says that a manifold described by a collection of patches or charts has an associated family of functions f_α using which reduces the integration problems to a sum of integrals on the chosen local coordinate charts. Let us explain how this works.

3.4.1 *Partition of Unity*

Introducing a partition of unity is always possible for a certain kind of manifold. Thus a partition of unity introduces a family of functions f_α satisfying the following conditions

(1) f_α are smooth,
(2) $f_\alpha \neq 0$ in $U_\alpha \Rightarrow$ compact support, where U_α are locally finite cover of the manifold. This property is holds for manifolds that are paracompact and locally compactness which means that each f_α vanishes outside and on the boundary of U_α while paracompactness of the manifold means that any given point of the manifold belongs to only a finite number of coordinate neighbourhoods U_α.
(3) $0 < f_\alpha < 1$,
(4) $\sum_\alpha f_\alpha = 1$.

Now,

$$\int_{\mathcal{M}} d\omega^{(n-1)} = \int_{\mathcal{M}} (d\omega^{(n-1)}) \sum_\alpha f_\alpha,$$
$$= \sum_\alpha \int_{\mathcal{M} \cap U_\alpha} f_\alpha d\omega^{(n-1)},$$
$$= \sum_\alpha \int_{\mathcal{M} \cap U_\alpha} d(f_\alpha \omega^{(n-1)})$$

Once we have this expression we can just look at a specific coordinate chart which has or does not have a boundary. From this the Stoke's theorem follows. Stokes theorem can be written in the following suggestive way

$$\langle \mathcal{M}, d\omega \rangle = \langle \partial \mathcal{M}, \omega \rangle$$
$$\text{where } \langle \mathcal{M}, d\omega \rangle = \int_{\mathcal{M}} d\omega \ , \ \langle \partial \mathcal{M}, \omega \rangle = \int_{\partial \mathcal{M}} \omega.$$

This way of stating the Stoke's theorem suggests that the d-operator which is a local differential operator is in some sense the "adjoint" of the geometrical non-local boundary. This is indeed true and leads to the construction of Homology Groups where a manifold is constructed by gluing together, for example, simple Euclidean space objects called simplicies which have dimensions ranging from zero, points, lines, triangles, tetrahedrons and so on. The highest dimensional simplex will have the dimension of the manifold. We will not develop this theme but move on to a brief discussion of Homotopy Groups. Homology groups were constructed by Poincare and used to extend Euler's genus formula to higher dimensions.

One final comment: Stoke's theorem was written in the form of a pairing between a form ω and a manifold M that gives a real number makes it clear that ω and M are dual objects with the action of M on ω given by $\int_M \omega$.

3.5 Homotopy and Cohomology Groups

The de Rham cohomology groups was introduced as an answer to the question: Is a closed form (i.e. a form $\omega^{(p)}$ with the property $d\omega^{(p)} = 0$) always exact (i.e. $\omega^{(p)} = d\eta^{(p-1)}$). The fact that the answer is no was used to distinguish the circle from the disc. Thus a method for detecting the presence of the hole around which had the circle as its boundary was found using differential forms. There are other ways of spotting holes on a surface. Another way is by using Homotopy groups which were invented by Poincare precisely for this purpose. Later the idea was generalised to detect holes of higher dimensions. The idea of a homotopy group is very intuitive. The objects used to construct these groups are loops and higher dimensional closed surfaces which start from a point of the manifold and can encircle a hole and thus detect their presence.

Let us sketch the approach for a two dimensional surface, M, with a hole in it. We can pick a point on the surface and draw closed loops. These loops will be of two types. There will be loops that circle the hole and those that do not circle the hole. Again for loops that circle the hole there will be ones which circle the hole an arbitrary number of times in either the clockwise or anticlockwise direction. The idea of homotopy is to introduce a notion of equivalence between two loops that can be smoothly deformed into each other. Two loops that can be deformed into each other will be said to be homotopic. Thus given a loop α which starts and finishes at a point x_0 of the surface there will be an equivalent class $[\alpha]$ of loops homotopic α. Now a loop is nothing but a smooth map of the circle, S^1 to the surface. Poincare defined a group operation on the equivalence class of loops by introducing a way combining two loops to form a product loop . The combination rule for loops was the group operation. Following such a procedure Poincare constructed the first homotopy group, also called the Fundamental group of Poincare, $\pi - 1(M, x_0)$.

The rule for joining loops was simple. Each loop was taken to be a map $\alpha_i(t)$ on the surface with the parameter t in the closed interval: $[0 \leq t \leq 1]$ chosen so that $\alpha_i(0) = \alpha_i(1) = x_0$, a fixed point on surface. The important point was that a loop was to be parametrised by a variable t which was an element of a closed unit interval in such a way that the beginning and end of the loop were the fixed point x_0. Poincare introduced a combination rule for loops in the following way: consider two loops that start and end at the same point x_0. Now make the first loop return to its starting point in half the time interval followed by an equally fast moving second loop so that at the end of one unit of time the two loops are traversed. Thus the combined loops forms one loop. Poincare went on to show that by using the idea of homotopic loops the combination rule for homotopic loops form a group. There is also a notion of two spaces that can be continuously deformed into each other being of the same homotopy type. A startling example of this is that n dimensional Euclidean space and the origin of this space are spaces of the same homotopy type. The continuous deformation which establishes this equivalence is the map $x \to tx$, where x is a point of the n dimensional space and t as a real number. For $t = 1$ we get the point x while for $t = 0$ we get the origin. This example makes it clear that homotopy related spaces need not have the same dimension. Let us write down mathematical expressions summarising what we have said.

A loop α in a space M is thus a continuous map from the closed unit interval $[0, 1]$ to M with ends points fixed at a point $x_0 \epsilon M$. We write

$$\alpha : [0, 1] \to M$$
$$\alpha(0) = \alpha(1) = x_0$$

Thus a loop has three properties. It is continuous, it starts and end at a fixed point of M and the parameter t used to describe its position in M varies in the closed interval $[0, 1]$.

Now if α, β are two loops in M we define the combination rule of

Poincare to form a new closed loop γ as follows:

$$\alpha : [0,1] \quad \rightarrow \quad M$$
$$\beta : [0,1] \quad \rightarrow \quad M$$
$$\gamma : [0,1] = \alpha.\beta$$

where for all the loops the beginning and end points are fixed at $x_0 \epsilon M$ i.e. $\alpha(0) = \alpha(1) = \beta(0) = \beta(1) = \gamma(0) = \gamma(1) = x_0$. The rule for "multiplying" or combining closed loops is,

$$\gamma(t) = \alpha(2t) \text{ for } [0 \leq t \leq \frac{1}{2}]$$
$$= \beta(1 - 2t) \text{ for } [\frac{1}{2} \leq t \leq 1]$$

The parameter factors $2t$ for α and $1 - 2t$ for β are introduced so that both α, β have parameters ranging over the interval $[0,1]$ even though t ranges over $[0, \frac{1}{2}]$, for loop α and over $[\frac{1}{2}, 1]$ for loop β. The two closed loops meet at x_0 when $t = \frac{1}{2}$. Finally we write down what is meant by two loops being homotopic. Two loops α, β in M are homotopic if they can be continuously deformed into each other i.e. there is a "homotopy" parameter s belonging to a closed interval $[0,1]$ such that for $s = 0$ we get the loop α and for $s = 1$ we get the loop β. In other words α_s interpolates between α, β as s varies. Here is how this can be written. We have a homotopy map $\alpha_s(t)$ which is a continuous map from $[0,1] \times [0,1]$ to M defined to have the property,

$$\alpha_s(t) : \rightarrow M$$
$$\alpha_0(t) = \alpha(t)$$
$$\alpha_1(t) = \beta(t)$$

with $\alpha_s(0) = \alpha_s(1) = x_0$. If two loops α, β can be deformed into each other in this way they are said to be homotopic and we write $\alpha \approx \beta$. The equivalence class of loops homotopic to α is written as $[\alpha]$. The multiplication rule between two equivalent class of loops is defined to be $[\alpha] * [\beta] = [\alpha.\beta]$ i.e. representative elements from each class are multiplied and then their equivalent class is taken. It can be proved that such a procedure is well defined i.e. it does not depend on the particular representative elements chosen to define multiplication. Using this idea of multiplying equivalent class of loops Poincare showed that they form a group. This was called the Fundamental group of the space M in which the loops were introduced. The group captured a topological property of M. It was the birth of algebraic topology.

There is a corresponding notion of two spaces X, Y being homotopic if X can be continuously deformed to Y.

Homotopy groups are important in physics because by using them it is possible to distinguish spaces which cannot be deformed into each other. There are, for instance, an increasing number of examples in condensed matter where there are defects or textures that which cannot be removed by continuous deformations. Such defects can be characterised by homotopy groups. Thus line defects are classified by π_1 while more intricate defects can be classified by higher homotopy groups. We will later show that the presence of Dirac points on a topological insulator can also be understood using homotopy groups.

For the surface with a hole it is intuitively reasonable that the surface can be deformed to a circle S^1 and our discussion suggests that $\pi_1(S^1, x_0) = Z$, where the integer Z describes the winding number, clockwise and anticlockwise of the loop round S^1. If we have $\pi_1(S^1) = n$ it means we have a loop of winding number n.

The higher dimensional Homotopy group $\pi_n(M, x_0)$ come from smooth maps of S^n to M. The cohomology groups $\mathcal{H}^n(S^m)$ are known for all values of n, m but this is not true for the homotopy groups $Pi_n(S^m, x_0)$. They are not known for arbitrary integers n, m. But many cases are known and two important cases that are known are $\pi_n(S^n, x_0) = Z$ and $\pi^3(S^2, x_0) = Z$. The second example is very interesting as it shows the power of the homotopy group to spot global features. In this case a twisting of S'^2 and S^1 to form S^3, that is not detected by the corresponding cohomology group $\mathcal{H}^3(S^2)$ which is the trivial group, 0.

There are tables listing homotopy groups. Thus one need not know how to calculate homotopy groups but only to understand how and where they can appear in a physics problem. This requires spotting that in a given situation there is a map from, say S^n to a given space X of interest. If this happens then the group $\pi^n(X)$ will be relevant. Such maps show very often up as boundary conditions. If the group $\pi^n(X)$ is non trivial it means the system has topological features which will imply physical consequences. Thus homotopy groups are important for physics. We list a few homotopy groups, which have appeared in physics problems, in the form of a table.

Let us make a few more comments. Defects represent a lack of smoothness in the value of a physical parameter. For instance if spin directions on a plane all point in the outward normal direction of a circle on the plane then the spin directions cannot be continuously assigned in the disc which has the circle as its boundary. There must be a a lack of continuity on the

Homotopy Group	S^1	S^2	S^3	S^4	Comment
π_1	Z	0	0	0	Winding Number
π_2	0	Z	0	0	Point Defect
π_3	0	Z	Z	0	Hopf fibration
π_4	0	0	Z_2	0	Witten Anomaly

disc. Defects, i.e. lack of smoothness of some physical feature, appear in many physical systems.

The mathematical way of describing such defects uses two spaces. The first space is physical space. The second space is the order parameter space of the system of interest. The order parameter space is attached to each point of physical space. For example if we have a two dimensional spin system the physical space is two dimensions while the two dimensional spin information is described by points on a circle S^1 that give a direction and unit magnitude for the spins in two dimensions.

Now consider a closed curve in this two dimensional space. At each point there has to be a unit direction assignment which means choosing a point of S^1. This procedure is can be described as a mapping of S^1, the spin directions to S^1, points on the closed curve. This map represents an element of $\pi_1(S^1)$. If the element is non trivial then there must be a defect inside the curve. This follows from a topological theorem which says a map $f : S^1 \to S^1$ cannot be smoothly continued from boundary S^1 to points inside the boundary i.e. points of the disc which has S^1 as its boundary. A generalisation of this theorem for $S^n \to S^n$ holds which says a non trivial map from S^n to S^n cannot be smoothly continued to the n+1 dimensional ball that has S^n as its boundary. The lack of smoothness signals the presence of a defect. In the cases described the defects required by topology are point defects.

Hopf fibration, is related to the fact that $\pi_3(S^2) = Z$. It is an unexpected topological result that shows up in many physical situations. The fact that $\pi_3(S^2) = Z$ means that a three sphere S^3 can be wrapped round a two sphere S^2 in a non trivial way. Let us try to understand how this can happen. One way to do this is by using the result that locally there are tow coordinate charts in each one of which a three sphere S^3 can be written as $S^1 \times S^2$. We will explain this result in the example of a Hopf fibration that we give. Once we have a representation of S^3 in terms of $S^1 \times S^2$ we can then understand how the S^2, present in such a description, can be wrapped round S^2 for each coordinate chart. But when this done it turns out that the S^1 present in the two charts get linked during the mapping.

This rather exotic structure is actually present in hidden form when we construct the superposition of two normalised wavefunctions ψ_1, ψ_2. We show that lurking in this simple example there is a decomposition of S^3 into $S^1 \times S^2$. Let us explain how. We have

$$\Psi = \alpha\psi_1 + \beta\psi_2$$

where α, β are two complex numbers which satisfy $|\alpha|^2 + |\beta|^2 = 1$. Writing $\alpha = x_1 + ix_2, \beta = x_3 + ix_4$ we see that for given ψ_1, ψ_2 the state Ψ is fixed by α, β that satisfy the normalisation condition. But this means $(x^2)_1 + (x^2)_2 + (x^2)_3 + (x^2)_4 = 1$ which is a point on S^3. However we can factor out one of phases in the superposition expression. Suppose we factor out the phase of α so that becomes real. Then the normalisation condition changes, as now $\alpha = x_1$, it has no imaginary part and we have $x_1^2 + x_3^2 + x_4^2 = 1$ which now represents a point on S^2 and we have the phase of α multiplying this, which is S^1. This simple example shows how S^3 is locally $S^2 \times S^1$. Remember that S^1 is a unit radius circle which can be written as $e^{i\theta}$. The description is local because the procedure described will not work when $|\alpha = 0|$. When this happens we need to factor out the phase of β rather than that of α. Thus representing the superposition of two wavefunctions has an exotic topological feature. The proper way to understand these matters is in the geometrical setting of fibre bundles. This setting is necessary to describe K groups.

The Homotopy group $\pi_4(S^3) = Z_2$ was used by Witten to argue that $SU(2)$ gauge theory was inconsistent. He showed, using this homotopy group result, that a path integral formulation gave zero as the value of the field theory partition function.

3.6   Fibre Bundles and Vector Bundles

We are now ready to take one more step in the mathematical study of spaces by examining the important idea of fibre bundles and vector bundle. We have explained at the beginning of the chapter how fibre bundles are natural objects present in the description of most physical systems. The fibre bundle always involve two spaces, a base space which is a manifold, and a fibre space that describes the physical system. For instance the base space could be the surface of the earth and the fibre space three dimensional euclidean space that represents wind velocity. However the precise mathematical description become important only when it tells that the gluing together of the two spaces gives rise to a space that is twisted or has holes in it. The presence of such features lead to novel physical behaviour. For example the presence of such a twist is responsible for the gapless states of the topological insulator. To unravel the topology involved we need to introduce a number of ideas. Using them we can define the K groups that can classify vector bundles. We will learn how K groups can be calculated for a vector bundle. Once these technical details have been explained we will use K groups to discuss the topological insulator. The procedure that we describe allows us to calculate K groups for a wide class of manifolds. A non zero K group has physical implications.

K groups and K theory are called generalized cohomology theories. This means that they satisfy most but not all the rules that abstractly describe cohomology groups of topology. K groups are specially constructed to classify vector bundles. Wave functions on a Brillouin zone are vector bundles. Vector bundles are special fibre bundles.

Let us quickly summarise basic features of fibre bundles. Fibre bundles are spaces built by gluing together two spaces: A base space B (in our case the Brillouin zone) and a fibre space F which is a vector space (in our case the space of wavefunctions). For a vector bundle there is a group G that acts linearly and transitively on fibre vector space i.e. it moves points round without having fixed points and any two points in the fibre space can be connected by a group operation. The group G is called the group of the fibre.

The gluing procedure is described by first representing the base space B by a collection of contractible overlapping open sets $U_\alpha, U_\beta, \dots$ so that on each open set U_α the bundle E is simply the product of the spaces U_α and F. The map from E to this product space is a map ϕ_α which describes the bundle E locally. If two open sets U_α and U_β overlap then the bundle E will have two coordinate descriptions in the overlap region. A map linking these two descriptions is required for consistency. This map $g_{\alpha|\beta}$ is called a transition function. It is the crucial step in the construction of a fibre bundle as it glues together untwisted spaces and can do so as to generates the twists and holes.

Finally, a bundle E has a projection map $\pi : E \to B$. It describes the way in which overlapping descriptions of E must be glued together to form the global bundle space E. The local product structure of the bundle E is called a "local trivialization".

In any one of these local trivializations, a point in the bundle can be described by a pair of points $(x, f(x))$ where x is a base point and $f(x)$ the corresponding location of the point on the fibre. This is thus a map from the base space to the bundle space. As x ranges over the base we get $f(x)$ ranging over the different fibres. The collection $f(x)$ is called a section of the bundle. It is in general not a function as different open sets are needed and after a global tour on the base when one returns to the same point x one need not return to the same $f(x)$ but to a rotated $gf(x)$ where $g\epsilon G$ the group of the fibre. Sections represent physical information. For instance they can be wavefunctions. If the bundle E can be described as the product of B and F then the bundle is said to be a trivial bundle. In this case the section becomes a function. An example would be when B is a circle S^1 and the fibre is a line L that are glued together to form a cylinder (a trivial bundle) while if the gluing is done with a twist we get the Mobius strip (a non trivial bundle). If the fibre space F is a vector space then the bundle is said to be a vector bundle. If the dimension of the fibre vector space is k then we have a fibre bundle of rank k. This is the basic intuitive idea of a vector bundle. Different ways of gluing lead to bundles that are not equivalent. Thus finding out all possible bundles given, the base B, the fibre F and the group of the fibre G will depend on the way the base space is represented by open sets and then how these are glued together. This is the problem of classifying a bundle.

There is a standard procedure for classifying vector bundles which reduces the classification problem to that of determining the homotopy class of maps from the base space B of the bundle of interest to the base space of a space called the classifying space BG. Fortunately we do not need to know the details of the space BG. All we will need for our calculations is G, which is the group of the fibre. Thus this result from fibre bundle theory tells us that bundles, with S^n as base space, are classified by the homotopy group $\pi_{n-1}(G)$. The point to note is that this result requires us to calculate the homotopy group not for a space but for a group. We will briefly explain how this is done by replacing the group by a space associated with it. Thus we will show that the groups $U(1), SU(2), SO(3)$ can be replaced by spaces S^1, S^3, RP^3 respectively.

Classifying vector bundles when the base space M is not an n- sphere S^n needs special methods as the classifying theorem requires finding the homotopy equivalent class of maps from $M \rightarrow G$. These are not standard homotopy groups. This is the problem that has to be solved for the topological insulator.

Before we describe the technical aspects of K theory we give an intuitive picture explaining why time reversal invariance can have topological consequences. Recall that the operation of time reversal on a spinor wavefunction in momentum space does three things: it flips the spin of a spinor wavefunction, it changes the momentum direction from k to $-k$ and it replaces the wavefunction by its complex conjugate. Consider a time reversal invariant system with a real wavefunction. The point $k = 0$ for such a system is energy degenerate for its spin-up and spin-down states. This means that if we consider a loop in momentum space which starts and ends at $k = 0$ this energy degeneracy impies that at the end of the loop there is the possibility for the spin-up and spin-down states to be identified. If this happens then a twist in the wavefunction occurs. The consequence of this is that the momentum space loop now has a twist that cannot be removed by continuous deformations. It is like a Mobius strip and is thus topologically stable. The topological twist comes from time reversal invariance and energy degeneracy at the point $k = 0$ in momentum space.

Topological arguments can also be used to show that the associated spinor wavefunction must have a gapless point. The intuitive explanation for a time reversal invariant system given can only suggests that such a system could have interesting topological features but does not explain why a time reversal invariant system, like the topological insulator, has a gapless Dirac states or it should have strong spin-orbit couplings. These issues are clarified by the K Theory calculation.

3.7   K Theory

To actually classify vector bundles, using K theory, for a large class of base spaces, we just need to know how to use two basic theorems of James and how to look up a table of Homotopy groups of spheres. Let us explain how this is done. We start by introducing a few technical terms that we need.

K Theory introduces an abelian group operation between bundles that comes from adding them. Two bundles are added by adding their fibres, which are vector spaces. This notion of adding bundles is restricted to bundles which have the same base space.

If bundles can be added then a simple trick allows us to define a notion of subtraction between bundles. Once this is done we have an abelian group whose objects are the bundle and the group operation is addition (with subtraction as the inverse operation). This is the K group of the bundle. Thus adding two vector bundles is simply to take the sum of their fibres as the new fibre. The two bundles must have the same base space. The usefulness of the K group comes, as we will show, from the key theorem:

Theorem

Given a vector bundle [E] of rank k with fibre of dimension k on a manifold of dimension n for $k > n$ it is possible to find an isomorphism between the bundle [E] and a bundle $[F^m] + [I^{k-m}]$ with $m < k$ where $[F^m]$ is a bundle of rank m and $[I^{k-m}]$ is a trivial bundle of rank $(k - m)$.

This means that a lower rank bundle $[F]^m$ contains all topological information present in [E]. Once a certain fibre rank is crossed no further topological information appears. This is called the stable range. We thus have a notion of the bundle equivalence:

Definition

Two bundles are equivalent if $[E] + [I^j] = [F] + [I^l]$, where two bundles [A],[B] are added to form the bundle [A]+[B] by adding their fibres (which are vector spaces). We write $[E] \sim [F]$. We have a convenient Lemma

Lemma

If [E]+[G]=[F]+[G], then $[E] \sim [F]$.

The proof is simple. For a reasonable base manifold it is always possible to find for a given [G] a bundle [G'] such that $[G] + [G'] = [I]$, where [I] is a trivial bundle.

Thus [E]+[G]+[G']=[F]+[G]+[G']=[E]+[I]=[F]+[I]
which means that $[E] \sim [F]$

3.7.1 *Subtracting Bundles*

We have described how bundles are to be added. We now explain how they can be subtracted. We explain the idea using real numbers. The fundamental idea used is that a number can be written in many different ways. Thus the result of subtracting two numbers can also be written in many different ways. Thus $7 - 5 = 8 - 6 = 9 - 7 = ...$ We exploit this idea and write,

$$s_1 - t_1 = s_2 - t_2 \tag{3.1}$$

as two ways of writing the same number. This implies

$$s_1 + t_2 = s_2 + t_1 \tag{3.2}$$

or as

$$s_1 + t_2 + u = s_2 + t_1 + u \tag{3.3}$$

where u is any arbitrary real number. This equation is a way of defining subtraction using addition.

We now use this idea to define the subtraction between two bundles. Thus $([E], [F]) = [E] - [F] = [G] - [H] = ([G], [H])$ means that $[E] + [H] + [J] = [G] + [F] + [J]$. We define the virtual dimension for the pair $([E], [F])$ as the difference between the ranks of the bundles $[E]$ and $[F]$. (The rank of a vector bundle is the dimension of the fibre vector space of the bundle). Note that the virtual dimension can be negative; it is zero if the bundle pairs have the same rank.

This notion of addition and subtraction between pairs of bundles introduced gives an abelian group structure which is the K group. Furthermore, for stable bundles of the same rank, we have, from the theorem quoted, an isomorphism between bundles. This is the case for Reduced K Theory where we only consider bundles of the same rank. If this restriction is not used, we have K Theory.

We now turn to the way the K groups can be calculated. The groups depend on the group of the fibre. For K Theory the fibres can have arbitrary dimension. We will suppose that the vector bundles are real vector bundles. Then the appropriate fibre group is the group $SO(k)$. For complex bundles, the group is $SU(k)$.

3.8   Calculating K Groups

In order to calculate K groups, it is convenient to introduce a pair of spaces and use them to construct three new spaces. The interrelationship between these spaces will give us a way to calculate the K groups. The reason for considering this collection of spaces is to tackle the problem of a fixed point from which loops of homotopy theory can begin and end. When we have a base manifold $M = X \times Y$ there is no obvious way of relating a starting fixed point for homotopy loops for this space to those of X and Y. The spaces introduced, together with maps linking them, solve this problem.

The spaces introduced are:

$X \times Y$, the tensor product of the two spaces,

$X \vee Y$, the disjoint union of the two spaces with one common point (the base point),

$X \wedge Y$, the smash product between the two spaces, which is the quotient space $\frac{X \times Y}{X \vee Y}$. We now explain the geometrical meaning of the smash product by a key example.

Suppose $X = S^1$, $Y = S^1$ then

$S^1 \times S^1$ is a two torus

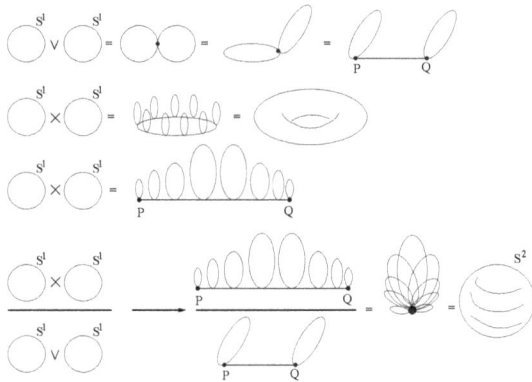

Fig. 3.4   Pictures for the operations introduced for two circles.

$S^1 \vee S^1$ is two circles with one common point (a figure eight)

$S^1 \wedge S^1$ is the quotient space $\frac{S^1 \times S^1}{S^1 \vee S^1}$.

We claim that this space is simply S^2. To see this we start by opening up one of the circles of the space $S^1 \vee S^1$ and think of the space as a line with end points P,Q each with a circle. The points P,Q are to be identified to get back to our original space.

We now think of the space $S^1 x S^1$ as a line with again end points P,Q, with circles at each point. Again identifying the end points together with the circles at those points gives back the starting space.

We can now describe the quotient space using these pictures (see Fig. 3.4)

Quotienting means taking the picture for $S^1 x S^1$ described and shrinking the line introduced together with its end point circles to a point.

We can think of the picture for $S^1 \times S^1$ in topological terms as circles of radii which are small at the two ends and reach a maximum radius at the middle. By shrinking the base line and the two end point circles we get S^2 (see Fig. 3.4). Shrinking the base line and the end points represents quotienting by $S^1 \vee S^1$.

3.8.1   *Exact Sequences*

The spaces introduced encode geometrical information. This information convert to maps between the associated K groups. These maps, relating the K groups, form a special collection of groups and homomorphisms called an exact sequence. Let us explain what exact sequences are.

Suppose we have two abelian groups A and B and a homoporphism between them. Recall that a homomorphism is a map $h : A \to B$ with the property $h(a_1)h(a_2) = h(a_3)$ in the group B if $a_1.a_2 = a_3$ in the group A. For a general homomorphism the map h need not be a $1-1$ map or even include all the group elements of B. Let us look at two cases. Consider the two groups $A = (1, -1, i, -i)$ and $B = (1, -1)$ with multiplication as the group operation. Introduce the homomorphism h given by,

$$\pm 1 \to 1$$
$$\pm i \to 1$$

In this case more than one element of the group A maps into the identity element of B. The elements of A that map into the identity element of B are called the kernel of the homomorphism, while the elements of B reached by h are called the image of the homomorphism h. In this case the kernel has two elements i.e. ± 1 of A is the kernel h while the two elements $pm1$ of B are the image h. Two groups are isomorphic if the image h is all of B and if kernel h has just one element. This just means that the two groups have the same number of elements and the way they combine is the same.

In the second case we let all the elements of A map into the identity element of B. Then the map does not include all the elements of B.

An exact sequence of homomorphisms between abelian groups, arranged in a linear manner, means that the image of a homomorphism, say $h : A \to B$ is the kernel of the homomorphism $g : B \to C$ that follows. For easy visualisation of such a sequence of groups and homomorphisms a diagrammatic representation is used. Let us look at a simple example of an exact sequence and show how, in this case, by analysing consequences of exactness the isomorphism of two groups is established.

Consider four abelian groups $0, A, B, 0$ two of which are trivial and have just one element, the identity element, which we write as 0 while the two non trivial groups are A, B. Now suppose that these groups form the exact sequence. The linear links between groups is displayed as follows,

$$0 \to A \to B \to 0$$

We will show that the groups A and B are isomorphic by listing implications of exactness,

$0 \to A$ image has only one element

$A \to B$ exactness means kernel has one element

$B \to 0$ exactness means all elements of B reached

From the second statement it follows that the map between A and B is $1-1$ as its kernel has just one element while the final statement tells us that the map from A reaches all elements of B. Thus A and B are isomorphic.

Let us return to the spaces $X, Y, X, \vee Y, X \times Y, X \wedge Y$ introduced. We state, without giving details, how links between these spaces get converted to an exact sequence of links between K groups.

The starting point is the natural set of maps between these spaces that exist, for example, the reduced join $x \vee Y$ can be regarded as a subset of the Cartesian product $X \times Y$. We have maps,

$$X \vee Y \to X \times Y \to X \wedge Y \tag{3.4}$$

These special maps induce an exact sequence of maps between K groups but with the direction of maps reversed. Before writing down this result, which we do not prove, we briefly comment on our notation and explain why the direction of maps gets reversed. The reason is a basic property of "dual" spaces.

The K groups depend on the group of the fibre and the space. In our discussions we suppress details of the fibre group and simply write $K(X)$ for a space X. The K group element for a space give a group element, a real number. This is the definition of a duality operation between linear spaces. Thus the K group of a space X is the "dual" of X. It is a generalised cohomology group. The idea of a dual space is important and is present whenever, for example, scalar products are introduced. Thus given v a vector and v^* a dual vector the property that $v^*(v) = $ a scalar can be written as $v^*(v) = (v^*, v) = (v, v)$. Thus the scalar product is simply a description of the action of the dual vector on a vector. In this case the two vector spaces are the same and we have $v^* = v$ with the action of v^* on v defined by the scalar product.

We now prove one an important feature of such spaces.

Consider a linear map $\alpha : X \to Y$, between two vector spaces X, Y and write their dual vector spaces as X^*, Y^*. This means that given $x \epsilon X$ and $x^* \epsilon X^*$ we have $x^*(x) = $ a scalar. The scalar can be a real or complex number. We now show that there is an induced map α^* between X^* and Y^* of the form $\alpha^* : Y^* \to X^*$ i.e. the induced map goes in the opposite direction. To prove this result we write $x^*(x)$ as (x^*, x). Then we have

$$(y^*, \alpha x) = (\alpha^* y^*, x)$$
$$(\alpha^* y^*, x) = (x^*, x)$$

where $\alpha^* y^* = x^*$. Thus α^* maps a pont y^* to a point x^* i.e. it goes in the opposite direction as claimed and we understand why a result of cohomology theory tells us that given maps between spaces induces maps between K groups which are in reverse order. The fact that the groups are dual to the spaces explains the direction reversal not that the sequence of maps are exact. That has to be proved.

$$0 \rightarrow K(X \wedge Y) \rightarrow K(X \times Y) \rightarrow K(K \vee Y) \rightarrow 0 \qquad (3.5)$$

From this it follows that

$$K(X \times Y) = K(X \wedge Y) + K(X \vee Y) \qquad (3.6)$$

We now use a result of K theory which tells us that

$$K(X \vee Y) = K(X) + K(Y) \qquad (3.7)$$

Thus we have our important formula,

$$K(X \times Y) = K(X \wedge Y) + K(X) + K(Y) \qquad (3.8)$$

Using this formula, and two theorems of James, the K groups for many manifolds with different fibre groups can be easily calculated.

The K groups depend on the nature of the vector bundle that are considered. We will take the vector bundles to be real and the group of the fibre to be $SO(N)$ groups. For the topological insulator we will take the fibre group to be $SO(3)$. As K theory gives results for bundles of arbitrary rank we will need to introduce restrictions on the results we obtain in order to make sure that our results are appropriate for $SO(3)$. To do this we need a result of James and Thomas.

To calculate the K groups we need two further results. The first is the result is a theorem of James which we will state in the section where we calculate K groups for the Topological Insulator and the second result is:

$$K(S^n) = \pi_{n-1}(G) \qquad (3.9)$$

Where G is the group of the fibre, $\pi_{n-1}(G)$ is the $(n-1)$ th homotopy group of G, and $K(S^n)$ is the K group for the n-th sphere S^n.

We sketch the proof of this result. To study the topology of S^n it is convenient to represent S^n as the union of two contractible spaces A,B. These spaces overlap to form the topological space S^{n-1}. For example for S^2 the space A is the upper hemisphere and B the lower hemisphere and they intersect along the equator S^1. A,B are contractible spaces so that bundles on them are trivial namely $A \times F, B \times F$ where F is the fibre space on which the group G acts. The nature of bundles on S^n thus depends on the way the two trivial bundles on A, B are glued together. This depends on the homotopy properties of the map from the intersection region between A, B to G. But this is just $\pi_{n-1}(G)$.

3.9 Groups and their Manifolds

In the last section the result $K(S^n) = \pi_{n-1}(G)$ was stated, where G is the group of the fibre with $\pi_{n-1}(G)$ the $(n-1)$ th homotopy group of G. We now explain the way the groups $U(1), SU(2), SO(3)$ can be described by the spaces S^1, S^3, RP^3 respectively. Indeed all the classical Lie groups have spaces associated with them that completely capture their cohomological properties. For instance $SU(N) \to S^1 \times S^3 \times ... \times S^{2N+1}$. We start with $U(1)$.

U(1)

The group $U(1)$ is a unitary group which means that $U^\dagger = U^{-1}$ i.e. the adjoint is the inverse. For $U(1)$ the adjoint is the same as U^*, the complex conjugate as it is a 1×1 matrix. Thus we can write a general $U(1)$ group element as $g(\theta) = e^{i\theta}$ where θ is real. But we know from Euler's theorem that

$$e^{i\theta} = \cos\theta + i\sin\theta$$

Thus a general element is fixed by choosing a value for θ restricted to lie on a circle S^1 since the function representing $U(1)$ is a periodic function of θ and we have $0 \leq \theta \leq 2\pi$. Thus the space associated with the group $U(1)$ is S^1.

SU(2)

The group $SU(2)$ is a 2×2 unitary matrix which has determinant equal to one. The property $U^\dagger = U^{-1}$ is satisfied by any 2×2 matrix of the form,

$$\begin{pmatrix} a & b \\ -b^* & a^* \end{pmatrix}$$

with $|a|^2 + |b|^2 = 1$, the determinant condition. But writing $a = x_1 + ix_2$ and $b = x_3 + ix_4$ the condition becomes $(x_1)^2 + (x_2)^2 + (x_3)^2 + (x_4)^2 = 1$ which is a point on S^3. Thus each point on S^3 gives an allowed $SU(2)$ matrix and is its associated manifold. **SO(3)**

The group $SO(3)$ is an orthogonal 3×3 matrix that represents rotations in three dimensions. We determine the parameter space needed to describe a general rotation. We will see it is the space RP^3. To describe a rotation we need to fix the axis of rotation and prescribe the rotation angle. The axis of rotation can be fixed by the polar angles α, β. The angle θ can be taken to be the radius of the sphere. Thus α, β give an axis direction and θ gives the length along this axis. Since $-\pi \leq \theta \leq +\pi$ the radius of the parameter space sphere is π. But a rotation along an axis by an angle $\pm\pi$ give the same point. Thus the parameter space sphere of radius π has to have the additional property that the two ends of any axis through the centre of the sphere meeting the surface of the sphere must be identified as they represent the same rotation. This is the space RP^3.

3.10   Topological Insulators and their K Groups

A topological insulator is an insulator which has a gapless surface point. We will use the topological methods discussed to show that three ingredients are necessary for this to happen, namely the topology of the base manifold of the system, the presence of strong spin-orbit coupling and the time reversal invariance of the system. Spin-orbit and time reversal invariance changes the Pauli matrix group from $SU(2)$ to $\frac{SU(2)}{Z_2}$, topology comes from periodicity in momentum space which makes the base space Brilluoin zone a three torus: $S^1 \times S^1 \times S^1$. These features lead to gapless surface but not bulk states for sufficiently strong spin-orbit interactions. Let us give the details. We start with topology.

Topology comes from the lattice structure. For instance in the tight binding model the nearest neighbour distance scale translates into a periodic structure in the Brillouin zone which is thus a non-contractible base space with topological properties.

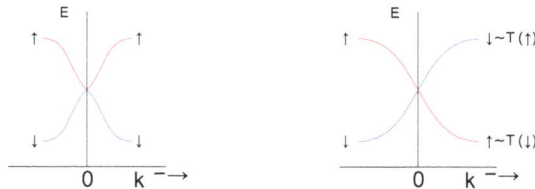

Fig. 3.5   Effect of time reversal on energy levels.

To understand the need for strong spin-orbit interaction and time reversal invariance it is best to examine an effective Hamiltonian describing an bound electrons, with no electrons in the conductance band where we neglect spin-orbit interactions. Thus,

$$H = \frac{\mathbf{p}^2}{2m} + V(r)$$

where $V(r)$ is a function which has lattice information in it. The fermion described by this Hamiltonian are spinor wavefunctions $\psi(x,t)$, with fibre group $SU(2)$ but the spin degrees of freedom play no role for this Hamiltonian. Under time reversal $\psi \to \psi^*$.

As soon as spin-orbit interaction is included the situation changes in two ways. Firstly if the spin-orbit interaction is strong enough surface state electrons can be pushed up to the conduction band and secondly the behaviour of the wavefunction under time reversal changes. Now under time reversal $\psi \to \sigma_2 \psi^*$. A consequence of this is that the fibre group changes from $SU(2)$ to $\frac{SU(2)}{Z_2}$. This change leads to a non zero K group only for surface states, not for the bulk and only when the fibre group is $\frac{SU(2)}{Z_2}$ not $SU(2)$. Finally a link between K groups and another topological invariant, the index theorem, implies that a non zero K group on the surface implies the presence of a gap less surface point. In the neighbourhood of this point a Dirac description of the system is appropriate. We now give the details of this arguments.

The change of the fibre group from $SU(2)$ to $\frac{SU(2)}{Z_2}$ is a key step of the topological argument sketched hence we briefly review how time reversal invariance leads to this result.

3.10.1 *Statement of Problem*

The surface properties of the topological insulator can be modeled by spin half electrons that are tightly bound, have strong spin-orbit interactions and are described in terms of a time reversal invariant Hamiltonian. A quasiparticle description for such a system near a conducting point is given by a mass less Dirac equation. Let us try to see how can this can happen as the surface electrons of the system are non relativistic particles.

Let us suppose that there is a gapless conducting point on the surface and construct an effective Hamiltonian valid near such a point. Let us list expected the features such an effective quasiparticle Hamiltonian H

(1) The Hamiltonian should be rotational invariant.
(2) The spin half nature of the quasiparticles means that we must include Pauli spin matrices in the Hamiltonian.

(3) Topological insulators also have strong spin-orbit coupling this requires that such an interaction term be included.

(4) The Hamiltonian must be time reversal invariant.

It is easily checked that a Hamiltonian, linear in momentum with mass and spin-orbit couplings has these properties.

3.10.2 *The Hamiltonian*

Consider the following Hamiltonian with a mass term and spin-orbit coupling in coordinate space.

$$H = \sigma.p + g\sigma.L + m\sigma_0$$

Here $\sigma_0 = I$, I is the identity matrix, while σ_i for $i = 1, 2.3$ are the Pauli matrices. L is the angular momentum vector and m the mass. We will show that such a Hamiltonian describes a time reversal invariant system. In two dimensions $\sigma.L = \sigma_3 l_3$

3.10.3 *Time Reversal Invariance*

The time reversal transformation reverses the direction of time, momentum, angular momentum and transforms the wavefunction. Thus we have,

$$t \to -t$$
$$x \to x$$
$$p \to -p$$
$$L \to -L$$
$$\sigma \to -\sigma$$
$$\psi \to \psi_T$$

The combination $\sigma.L$ is thus time reversal invariant. Time reversal invariance requires ψ_T to obeys the same equation as ψ but with t replaced by $-t$. The time reversed wavefunction thus evolves in exactly the same way as the original wavefunction but in the reverse time direction.

We want to determine the relationship between ψ and ψ_T. To find such a relationship we will need to use the following transformations under complex conjugation: $\sigma_1^* \to \sigma_1, \sigma_2^* \to -\sigma_2, \sigma_3^* \to \sigma_3$ and, since both p and L are imaginary in coordinate space, $p^* \to -p, L^* \to -L$.

Using these results in our Hamiltonian

$$H = \sigma.p + g\sigma.L + m\sigma_0$$

we find that,

$$\sigma_2 H^* = H\sigma_2$$

taking g, m real. The result is easily checked by a direct computation. We sketch the basic step for one term of the Hamiltonian. We have

$$H = (\sigma_1.p_1 + \sigma_2.p_2) + \dots$$
$$H^* = (-\sigma_1.p_1 + \sigma_2.p_2) + \dots$$
$$\sigma_2 H^* = H\sigma_2$$

where we used the transformation properties of the variables under complex conjugation as well as the commutation relations of the Pauli matrices. Now, setting $\hbar = 1$ the quantum evolution equation for ψ is,

$$-i\frac{\partial \psi(k,t)}{\partial t} = H\psi(k,t)$$

We want the time reversed wavefunction to satisfy the same equation but with $-t$ replacing t, namely,

$$-i\frac{\partial \psi_T(t)}{\partial t} = H\psi_T(t)$$

On the other hand by taking the complex conjugate of the equation for ψ and replacing t by $-t$ we get,

$$-i\frac{\partial \sigma_2\psi^*(-t)}{\partial t} = \sigma_2 H^*\psi^*(-t)$$

which gives

$$-i\frac{\partial \sigma_2\psi^*(-t)}{\partial t} = H\sigma_2\psi^*(-t)$$

where the result $\sigma_2 H^* = H\sigma_2$ established is used. Thus setting

$$\psi_T(t) = \sigma_2\psi^*(-t)$$

it follows that indeed $T(\psi, t) = \sigma_2\psi^*(-t)$ satisfies the same time evolution equation as ψ with t replaced by $-t$.

It is easily checked that the operator $T\psi(x,t) = \sigma_2\psi^*(x.-t)$ has the property that $T^2 = -1$. This means that the $SU(2)$ symmetry of the spin half quasiparticle is reduced to $SO(3) = SU(2)/Z_2$, where Z_2 is the group generated by T^2. This is the important result we need to carry out our K theory classification. We briefly note that:

(1) for spin zero wavefunctions the transformation $\psi \rightarrow \psi^*$ for any real Hamiltonian is time reversal invariant.

(2) From dimensional analysis we see that the coefficient of the linear term in momentum has a coefficient with the dimensions of speed (we have set this coefficient equal to one) and the mass term and spin-orbit terms must have coefficients so that they both have the dimensions of energy,

(3) In two dimensions an additional time invariant term is possible, namely $h\epsilon_{ab}\sigma_a p_b$, where h has the dimensions of speed. We note that the equation discussed is the two dimensional Dirac equation with spin-orbit put in by hand.

(4) In three dimensions the spin-orbit term $l.s = j^2 - l^2 - s^2$ is insensitive to the value of s_z i.e. whether the system has spin up or down. This is not the case in two dimensions where $l.s = l_z s_z$ and thus changes sign pushing up/down the energy depending on the sign of s_z for any fixed l_z value.

Thus we have constructed the general time reversal invariant Hamiltonian and explained how a Dirac equation emerges in the low momentum region, provided the system has a gapless point.

3.10.4   *Classifying bundles for the Topological Insulator*

To classify possible bundles for this system we thus need to calculate the real K theory groups for $SO(3)$ ie $K(S^1 \times S^1 \times S^1)$ with fibre group $SO(3)$. Using the results of the previous section we have:

$$K(S^1 \times Y) = K(S^1 \wedge Y) + K(Y) \text{ with } Y = S^1 \times S^1$$
$$K(Y) = K(S^1 \times S^1) = K(S^1 \wedge S^1) + K(S^1) + K(S^1)$$
$$K(S^1 \wedge S^1) = K(S^2)$$

Thus we get

$$K(S^1 \times S^1 \times S^1) = K(S^1 \wedge Y) + \pi_1(SO(3))$$
$$+\pi_0(SO(3)) + \pi_0(SO(3))$$
$$+\pi_0(SO(3)) \tag{3.10}$$

To calculate $K(S^1 \wedge Y)$, where $Y = S^1 \times S^1$ we need a theorem of James

Theorem

If the space Y has a set of Betti number $b_0, b_1, ...b_k$ each b_r term would lead to a factors $K(S^{r+1})$. The resulting K group would be a direct sum of such terms.

In our example $Y = S^1 \times S^1$ is a two dimensional spaces with Betti numbers $b_0 = 1, b_1 = 2, b_2 = 1$. Then from the theorem, it follows that $K(S^1 \wedge Y) = K(S^1) + K(S^2) + K(S^2) + K(S^3)$.

Using this result we finally get

$$K(S^1 \times S^1 \times S^1) = Z_2 + Z_2 + Z_2 + K(S^3) \qquad (3.11)$$

We have dropped the $\pi_0(SO(3))$ groups. We have retained $K(S^3)$ since this is not in the stable range (where the rank must be greater than the dimension of the base, while here the rank and dimension of the base are equal).

To deal with this problem we use a theorem James and Thomas. A new term, the Steifel Whitney class appears in the statement of the theorem. These classes are determined by cohomology groups with Z_2 coefficients and are listed in tables of cohomology groups.

Theorem

The map $[T^3, BSO(3)] \to [T^3, BSO] = K(T^3)$ is injective, and under this map the elements of $[T^3, BSO(3)]$ correspond to the subgroup of $K(T^3)$ with vanishing third Steifel Whitney class.

Here $T^3 = S^1 \times S^1 \times S^1$ and the K groups are the reduced K groups where contributions from points are dropped. Looking up the Steifel Whitney class $H^3(S^3, Z_2)$ we find it is non zero. It is in fact Z_2. Hence $K(T^3)$ is the trivial group. Thus our K theory analysis gives:

$$K(S^1 \times S^1 \times S^1) = Z_2 + Z_2 + Z_2 \qquad (3.12)$$

From topology we also can conclude that the system must have a gap-less Dirac point on the surface but there are no gap-less points in the bulk as $K(T^3)$ is the trivial group. These results from a direct link between gap-less states and the presence of non zero K groups. Let us make this link explicit.

3.10.5 *Dirac Points*

Topological insulators, as we have stressed, are systems that are invariant under time reversal symmetry and have strong spin-orbit coupling. The presence of spin-orbit interactions means that the Pauli spin matrices must appear. However time reversal symmetry requires, as we saw, the replacement of the group $SU(2)$ of the spin matrices by the group $SO(3)$.

The groups of K-theory were constructed using the fact that vector bundles could be added. Using addition a notion of "subtraction" was found leading to an Abelian group structure which we used to classify vector bundles of interest.

We now point out that there is a link between $K(X)$ and a topological invariant called the index theorem. The index theorem relates topological properties of a bundle to zero modes of differential operators that act on sections of the bundle. Let us explain how this can happen by looking at an example. Suppose we have a Hamiltonian

$$H = i\gamma_\mu.D_\mu$$

which is chiral invariant. For four dimensions this means that the operator $\gamma_5 = i\gamma_1\gamma_2\gamma_3\gamma_4$ commutes with H. The D_μ operator is of the form $D_\mu = \partial_\mu - A_\mu$. Invariance under chirality means that if $H\psi = E\psi$ then $\gamma_\mu D_\mu \psi$ also has energy eigenvalue E. But $\gamma_5\gamma_\mu D_\mu = -\gamma_\mu D_\mu$ which means that for a chiral symmetric system each eigenvalue E is doubly degenerate corresponding to the two eigenvalues $\pm$ of γ_5. However this argument fails for $E = 0$. If the number of eigenvalues of $\pm\gamma_5$ are $n\pm$ then the index of the operator is defined to be the difference of these two numbers. The index can be shown to be invariant under deformations that preserve chiral invariance. It is intuitively reasonable, for instance, that the index should not change under the scaling of the operator as scaling the operator has no effect on zero energy states. A field theory way of proving the index theorem is to calculate

$$\text{index } i\gamma_\mu D_\mu = \text{Trace}(\gamma_5 e^{-\beta H})$$

Intuitively, contributions to the trace coming from E nonzero cancel as the γ_5 factor makes them appear as pairs of degenerate states with opposite signs which cancel. Thus only the zero energy state contributes. The constant $\beta > 0$ can be arbitrary.

In this way we get, in two dimensions, the formula

$$\text{index } i\sigma.D = \frac{1}{2\pi}\int F$$

where $F = dA$ the field strength.

The link between $K(X)$ and the index theorem allows us to infer that a non trivial $K(X)$ means that must be a zero mode. Once this is known we can use our in linear momentum Hamiltonian which gives a Dirac equation.

We now sketch the chain of ideas that relate K theory to the index theorem. No attempt to explain these results or prove them is made. The argument is sketched because the link between K theory and index theory is such powerful way to establish the existence of a gapless point that some discussion of the result seems appropriate.

The first step used to relate $K(X)$ to the index theorem is the following mathematical link between $K(X)$ and Chern character $ch(X)$, which we define shortly:

$$ch : K(X) \to H^*(X, \mathbf{Q})$$
$$[E]\text{-}[F] \to ch[E]\text{-}ch[F]$$

It is possible to check that the map is well defined which means it does not depend on specific choices made for $[E], [F]$. Index theory appear because there is a Chern character-$K(X)$-index theory link. For two dimension surface it is very simple, namely

$$\text{index}(D) = \ker D - \operatorname{coker} D = ch[X] = K(X)$$

where the notation ch[X] means that the integral of the Chern character is to be carried out on the manifold X. Finally we define the Chern character $ch(X)$ as defined as ch[X] $= \int \text{Trace } e^{i\frac{F}{2\pi}}$, where the integral is over the manifold X and $F = dA$ with A a connection on the bundle.

Let us repeat again the important message: we found $K(X)$ to be non zero on the surface of X but not in the bulk. Since a non vanishing $K(X)$ implies a non vanishing index, this implies a non vanishing kernel or coker-nel for the relevant operator, which means that the system has a zero mode (gap less surface point). In the neighbourhood of this point we then have a two dimensional Dirac operator with a $\frac{SU(2)}{Z_2}$, connection because of time reversal invariance. i.e. the gap less point of a topological insulator is a Dirac point.

Thus K-theory combined with the index theorem tells us that there must be a gapless surface point for the topological insulator where the system is described by a mass less two dimensional Dirac equation with a $SO(3)$ connection. No other details of the system are required.

We now briefly explain the idea of a connection in an intuitive way. Remember a bundle is constructed by gluing together two spaces, a base space and a fibre space. In our case the base manifold is the Brillouin zone and the fibre space is an appropriately chosen complex space that represents the wave function. In our case the fibre space is the space of spinor wave functions. The spinor in the the language of fibre bundle

theory are called sections of the bundle. A differential equation in this
language is a map between two sections of the bundle. A connection is a
modification introduced to the differential operators so that they can map
sections located at different neighbouring points of the base manifold. It is
necessary because the base manifold could have curvature. In physics the
connection is the vector potential.

Let us now make these ideas precise. The connection is a structure
which makes it possible to compare the way a tangent plane of a curved
manifold changes. Thus we introduce the geometrical idea of breaking up
the bundle space into a horizontal and vertical part at a point p of the
bundle. Neighbouring tangent planes can be compared by noting that a
component of a the coordinate changes gets rotated as we move along a
curved manifold. The rate at which the vector X_j rotates as we move
along a curve with tangent vector X_a gives a measure of the curvature.
This is the geometric picture of a moving frame on the base manifold. The
amount of rotation introduced is controlled by a rotation matrix ω^i_j. The
matrix ω is the connection.

We now formulate this idea in terms of a decomposition of the bundle
space into a horizontal and vertical part. The point is because of curvature
the horizontal and vertical parts of a bundle get mixed.

Let us represent a point of the bundle by a pair (x, g) where x represents
a local Euclidean coordinate representation of a point on the base manifold
and g an element of the group of the fibre G. Thus the fibre is taken
to be the group G itself rather an a representative vector space on which
the group acts. Such a fibre bundle is called a Principal bundle. Let $\mathbf{X}$
represent the tangent vector at x to a curve in the bundle. Because the
base manifold is curved we have

$$\mathbf{X} = \alpha_{ij}\frac{\partial}{\partial g_{ij}} + \beta_\mu\left(\frac{\partial}{\partial x^\mu} + C_{\mu ij}\frac{\partial}{\partial g_{ij}}\right)$$

where

$$\frac{\partial}{\partial g_{ij}} \text{ and } \frac{\partial}{\partial \mu} + C_{\mu ij}\frac{\partial}{\partial g_{ij}}$$

are the basis vectors for the vertical and horizontal spaces respectively with
$\mu = 1, 2..n$, and $i, j1, 2, ...d$ assuming the horizontal space has dimension n
and the vertical space has dimension d. The connection ω is now written
as

$$\omega = g^{-1}dg + g^{-1}\mathbf{A}g$$

and $C_{\mu ij}$ can be related to $\mathbf{A}$ by requiring that $< \omega, \mathbf{X} >= 0$. That is, the connection ω only rotates vectors in the fibre as we move along the horizontal direction. Such a rotation is due to the curvature of the horizontal space. This condition gives $C_{\mu ij} = -A_\mu^a \frac{\lambda_{aik}}{2i} g_{kj}$, $\mathbf{A} = A_\mu^a (x)(\frac{\lambda_a}{2i})dx^\mu$, where $[\frac{\lambda_a}{2i}, \frac{\lambda_b}{2i}] = f_{abc}\frac{\lambda_c}{2i}$. The connection can be written in the more suggestive way as

$$\omega = g^{-1}(d + \mathbf{A})g$$

In this form $\mathbf{A}$ can be recognised to be the vector potential or the connection of general relativity and $d + \mathbf{A}$ as the covariant derivative $D = d + \mathbf{A}$. In this form it is clear that physical or geometrical information is contained in $\mathbf{A}$. For instance $D \wedge A = dA + A \wedge A = F$ is the field strength in gauge theory and is the curvature tensor in general relativity.

We have outlined a general method of calculating K groups for a general class of spaces and have illustrated the method for the space $S^1 \times S^1 \times S^1$ and group $SO(3)$. We have explained why the space and group are relevant for topological insulators. In carrying out the calculation we needed two special theorems. These theorems allowed us to reduce the calculation of K groups for spaces $S \wedge Y$ to the sum of K group for spheres. The spheres needed were, according to a mathematical result, were identified by the Betti numbers of the space Y. The mathematical result thus reduces the problem of calculating K groups for spaces of this kind to calculating K groups of spheres.

The theorems also allowed us to identify the subgroup of the K groups that corresponded to the fibre group $SO(3)$. Our results confirm show that such a restriction modifies expectations of topological groups based on simple arguments. Thus for topological insulators there is three Z_2 groups not four. The usefulness of K theory is that it allowed us to calculate topological properties for bundles on the space $S^1 \times S^1 \times S^1$ in a purely algebraic way.

We note that if our bundle group had been $SU(k)$ for any k, i.e. if we were dealing with complex bundles, then the resulting K groups are all trivial. Thus time reversal invariance which changes $SU(2)$ to $SO(3)$ is essential. Without time reversal symmetry there would be no topological reason for the stability of the topological insulators.

We also used topological reasoning to show that there is a single stable Dirac point. For multiple Dirac points our analysis tells us we need to construct the tensor product of the vector bundles. The corresponding K groups are then simply given by the product of the number of the individual

K groups present, one for each Dirac points that is present. This immediately gives the result that the K groups are trivial for an even number of Dirac points but are non trivial for an odd number of such points. The result follows from the fact that the product of two (or even number) of Z_2 terms is trivial but is not trivial for an odd number.

One final comment. We have considered a system with time reversal invariance and found it has important topological consequences. This would suggest that one should look at other discrete symmetries which are broken by interactions and see if they too lead to topological consequences. Wigner's Theorem explains how discrete symmetries can be represented for Lorentz invariant systems but it also points out that the representations constructed, for instance, for a system that has both parity (P) and time reversal (T) symmetry, are projective representations. Such representations have undetermined phase factors which cannot be fixed by the theory. This limitation becomes important when linear combinations of projective representations, in the form of wavefunctions, need to be considered. Thus if we are given a projective representation on a space which is the direct sum of, say, two one dimensional invariant subspaces then we are allowed only one overall phase factor for each group element in this direct sum space but not allowed to introduce separate phase factor for each of the one dimensional subspaces. Hence two such representations even on spaces of the same dimension need not be equivalent. In terms of a physics problem this means that a knowledge of projective representations of a symmetry group is insufficient when interaction between particles are included. Interactions involve direct sum of representations. The usual group theory driven selection rules for allowed events that follow from the properties of irreducible representations cannot be used as now we have a collection of inequivalent representations describing the system.

However this problem can be avoided by a mathematical procedure which replaces a projective representation by a non projective representations of a different group. This is an intriguing result as it suggests that the correct symmetry of the system with P, T symmetry is not the Lorentz group extended by these discrete symmetries but the Lorentz group extended by a different discrete group. The symmetry group of the system thus has to be changed if we would like to consider direct sum of representations in order to treat interactions using group representations. But the new discrete non projective groups allowed are not unique. Using any one of these extended groups is allowed by the mathematics. To determine which one, if any, of them is realised in nature we have to turn to experiment. These remarks make it clear that abstract mathematical reasoning can suggest new avenues of exploration to unravel unexpected consequences of symmetry.

3.11 Morse Theory and Symmetry Breaking

We now explain how Morse theory, a topological method for studying global features of maxima and minima of a real function on a manifold, can be used to study symmetry changes of a crystal. We first explain the physics problem.

Suppose we have a crystal described by a density function $\rho(\vec{x})$ invariant under one of the finite crystallographic groups, say G, This means that,

$$\rho(g\vec{x}) = \rho(\vec{x}) \quad \text{with} \quad g \in G , \tag{3.13}$$

where $\rho(\vec{x})dV$ is the probability to find an atom of the crystal in the volume dV. When external conditions are changed we suppose the crystals symmetry changes from G to H_i which is a subgroup of G. We further suppose that this symmetry change is a second order phase transitions. A simple but very fruitful way to study such transitions was established by Landau. We use Landau's ideas. Since the symmetry change is a second order phase transition the density function ρ is expected changes smoothly from one phase to the other, even though the symmetry group G suddenly changes to a subgroup $H \subset G$. The density function, near the transition, can then be decomposed as

$$\rho(\vec{x}) + \delta\rho(\vec{x}) \tag{3.14}$$

where ρ is G-symmetric, while $\delta\rho$ is H-symmetric, and, at temperatures below the critical value $T < T_c$, $\delta\rho = 0$, while for $T > T_c$ is $\delta\rho \neq 0$. As the function changes continuously during the transition, $\delta\rho$ is small near T_c. Such a transition is an examples of spontaneous symmetry breaking: where although the thermodynamic potential $\Phi(\rho, T, P)$ is G-invariant its minimum, ρ_0, need not have this property. This minimum represents the equilibrium ground state of the system and we will call it the "vacuum" state. Thus by *vacuum* we mean the ρ configuration that minimizes the functional $\Phi(\rho)$ for a given value of T and P and we can have a situation where for $T < T_c$ the vacuum is G-invariant, realizing an explicitly symmetric phase; while for $T > T_c$ the vacuum is H-invariant, representing a spontaneously broken symmetric phase. Hence given a G-symmetric crystal (ρ and Φ) determining which are the allowed sub-groups H_i for the broken symmetry is equivalent to finding out if a H_i invariant minimum of Φ, ρ_0 is possible.

Using physical reasonings Landau assumed that, near T_c, Φ is a polynomial in ρ of three terms of order zero, two and four – the Landau polynomial. As ρ is a G-invariant function it can be expanded in terms of the basis of the functions, the characters of the group G, expansion, $\{\phi_i(\vec{x})\}_{i=1,...,\mathrm{ord}(G)}$, where $\mathrm{ord}(G)$ is the number of elements (order) of the finite group G, and $\phi_i(\vec{x}) = \sum_{j=1}^{\mathrm{ord}(G)} D_{ij}[g]\phi_j(\vec{x})$, $i = 1, ..., \mathrm{ord}(G)$, for any $g \in G$. The matrices $\{D_{ij}\}$ realize a reducible representation of G and, as ρ, the density is real, they are necessarily orthogonal: $\delta\rho(\vec{x}) = \sum_{i=1}^{\mathrm{ord}(G)} \eta_i\phi_i(\vec{x})$, and

$$\delta\rho'(\vec{x}) = \delta\rho(g\vec{x}) = \sum_{i=1}^{\mathrm{ord}(G)} \eta'_i\phi_i(\vec{x}) , \qquad (3.15)$$

with $\eta'_i = \sum_j D_{ij}[g]\eta_j$. The η_is are the *order parameters* of the phase transition ($\eta_i = 0$ at $T < T_c$, $\eta_i \neq 0$ at $T > T_c$), hence the thermodynamic potential will be minimized with respect to these quantities $\Phi(\vec{\eta})$. Landau suggested that the problem could be further simplified by restricting the expansion to just irreducible representations.

The reducible set $\{\eta_i\}$ can be decomposed into irreducible sub-sets $\{\eta_i^a\}$ where $\eta_i^a = \sum_{b=1}^n d_{ab}[g]\eta_i^b$, $a = 1, ..., n$, the $d_{ab}[g]$s being the $n \times n$ irreducible matrix-blocks in the $\mathrm{ord}(G) \times \mathrm{ord}(G)$ matrices $D_{ij}[g]$ and, supposing there are m such irreducible representations of dimension $n_1, n_2, ..., n_m$, we have

$$n_1^2 + n_2^2 + ... + n_m^2 = \mathrm{ord}(G) . \qquad (3.16)$$

Note that $\{\eta_i^a\}$ (or the corresponding $\{\phi_i^a\}$) is not a complete set. Nonetheless, it is only one such set that is to be used to expand $\delta\rho$ near T_c. Thus, the problem of finding which way the G-symmetry of the crystal is spontaneously broken down to the smaller symmetry of one of its subgroups – say H – boils down to the minimization of the real function(al)

$$\Phi : \vec{\eta} \in S^n \to \Phi(\vec{\eta}) \in R \qquad (3.17)$$

where n is one of the m values in (3.16), and we added to R^n the point at infinity that, for stability, has to be included as a maximum, so that $R^n + \{\infty\} = S^n$. In what follows we shall take $n = 3$.

One way of solving this problem is, of course, to explicitly know the coefficients of the Landau polynomial that depend upon the details of the model. Many things, though, can be said about the critical points (minima, maxima, saddle points) of a real function like Φ in (3.17), *without knowing its explicit expression*, but rather exploiting the constraints associated with the topology of S^n, as shown in the mathematical work of Morse. To do this two steps are required. We need to state the topological constraints on critical points on a manifold and then we need to show how the number of critical minima, with symmetry, is related to Ord$H$.

We first outline the topological result of Morse. In a nutshell, Morse proved that, when a function is like Φ, i.e. smooth, real, with isolated critical points and defined over a compact differentiable manifold (as said, we shall focus on the case of S^3), then the following constraints, known as Morse Inequalities, hold. In order to write these constraints we need to define a critical point and the index of a critical point. Suppose $\Phi(x)$ is a non constant real valued function on a manifold M of dimension n. Thus the point $x\epsilon M = (x_1, x_2, ...x_n)$ in terms of a local coordinate description of M. A critical point $(a) = (a_1, a_2, ...a_n)$ is where

$$[\frac{\partial \Phi}{\partial x_i}]_{x=a} = 0$$

Let us look at a particular critical point and choose coordinates so that $(a) = (0)$. Morse proved that if the critical point was isolated and non-degenerate the function $\Phi(x)$ could be approximated near the critical point by a quadratic function of the form,

$$\Phi(y) \approx -(y_1)^2 - (y_2)^2... - (y_k)^2 + (y_{k+1})^2 + ...(y_n)^2.$$

i.e. it has $-k$ negative and $n - k$ positive terms. We have set $\Phi(0) = 0$. Geometrically the picture is that there were k directions in which the value of Φ decreases and $n - k$ remaining directions in which it increases. Thus each critical point is described by its location and by its associated k value. The number k is called the index of the critical point. Both the locations and the values of k depend on Φ. What Morse found was that there were topological constraints on the number of critical points which he expressed in the form of a set of inequalities. For the manifold S^3 they are,

$$c_0 \geq b_0 \qquad (3.18)$$

$$c_1 - c_0 \geq b_1 - b_0 \qquad (3.19)$$

$$c_2 - c_1 + c_0 \geq b_2 - b_1 + b_0 \qquad (3.20)$$

$$c_3 - c_2 + c_1 - c_0 = b_3 - b_2 + b_1 - b_0 \qquad (3.21)$$

where c_ℓ is the number of critical points of Φ with index ℓ ($\ell = 0$ are the minima, $\ell = 3$ the maxima, $0 < \ell < 3$ the saddle points), and the b_ℓs are the Betti numbers of S^3 ($b_0(S^3) = 1 = b_3(S^3)$), while the other Betti numbers of S^3 are all zero). This is a powerful result: without knowing the actual form of the function, the topology of the manifold tells us that for each ℓ there are at least b_ℓ critical points with index ℓ: $c_\ell \geq b_\ell$. For the case in point we know that Φ has at least one minimum ($b_0 = 1$) and at least one maximum ($b_3 = 1$).

The topological Morse constraint has now to be related to group theory. This is done by using the fact that if a critical point lowers the symmetry from G to a subgroup H_i then the number of such critical points is fixed by group theory to the ratio $\dfrac{\text{Ord } G}{\text{Ord } H_i}$. Thus denoting by $\vec{\eta}^{(\ell)} \in S^3$ the critical point of Φ with index ℓ, and by $H^{(\ell)}$ the subgroup of G that leaves it invariant, $\eta_a^{(\ell)} = \sum_{b=1}^{3} d_{ab}[h]\eta_b^{(\ell)}$, $h \in H^{(\ell)}$, the number of elements (order) of $H^{(\ell)}$ is simply related to c_ℓ

$$\text{ord}(H^{(\ell)}) = \text{ord}(G)/c_\ell \ . \tag{3.22}$$

It is a result of the theory of finite groups that $\text{ord}(H^{(\ell)})$ as written in is indeed an integer, the proof being based on the coset decomposition of G with respect to $H^{(\ell)}$. The subgroup we are looking for is $H^{(0)}$, the symmetry of the vacuum configuration $\vec{\eta}^{(0)}$ and new symmetry H of the crystal beyond T_c. The constraints the requirement to have *at least* two maxima (at $\vec{\eta}^{(3)} = \infty$ and at $\vec{\eta}^{(3)} = 0$), i.e. $c_3 \geq 2$, and the fact that the number of *real* solutions of $d\Phi/d\vec{\eta} = 0$ (with $\vec{\eta} \in S^3$ and Φ a fourth order polynomial) is bounded to be

$$c_0 + c_1 + c_2 + c_3 \leq 3^3 + 1 = 28 \ , \tag{3.23}$$

prove to be enough to identify the permitted $H^{(0)}$s!

Which one is the actual H of the crystal after the spontaneous breaking of symmetry $G \hookrightarrow H$, of course, depends upon the particular physical situation under investigation. Examples are crystalline structures, such as the alloy Cu-Zn, that exhibit the octahedral symmetry, whose group O_h has order 48. The subgroups of O_h allowed by the requirement to be crystallographic groups, have order 8, 6, 4 and 2. Carrying on the analysis above outlined tells us that the only subgroups allowed are those for which: $\text{ord}(H^{(0)}) = 6$, which corresponds to C_{3v}, and $\text{ord}(H^{(0)}) = 8$, which corresponds to C_{4v} (recall that $\text{ord}(C_n) = n$ and $\text{ord}(C_{nv}) = 2n$). This means that C_{3v} and C_{4v} are *selected* as the only possible symmetries of a crystal, originally O_h-symmetric, undergoing a second order phase transition.

References and Further Reading

The Encyclopedia of Mathematics (MIT press), has tables cf homotopy, homology groups.

C. Nash and S. Sen, *Topology and Geometry for Physicists*, Dover. Basic ideas of topology and Morse theory are discussed.

C. Nash, *Differential Topology and Quantum Filed Theory*, Academic Press (2003). More advanced but readable. Gives background details needed to understand results quoted in the chapter.

C. Nash, Commun. Math. Phys. 88, 319(1983) Example of a K theory calculation

C. Nash and S. Sen, Reports on Mathematical Physics, **28**. 1 (1986). Example of a classification problem without using K theory

I.M. James and E. Thomas, J. Maths. Mech. **14**, 485 (1965). Basic theorem used in the chapter is proved.

I.M. James, Transactions of the AMS, **84**, 545 (1957). Proves a basic theorem used in the chapter.

S. Sternberg, *Group Theory and physics*, Cambridge University Press (1994) Discusses group theory in a geometrical way and explains how projective representations of a group can be replaced by a non projective representation of an extended group.

N. D. Mermin, Rev. Mod. Phys. **51**, 591 (1979) gives an intuitive discussion of topology and defects.

L. Michel, Rev. Mod. Phys. **52**, 617 (1980) contains a more mathematical exposition on defects.

R. Bott and L. W. Tu, Differential Forms in ALgebraic Topology, Springer-Verlag, New York (1982).

Supersymmetry and Morse theory, Edward Witten, J. Diff. Geom. 17 (1982) 661-692.

K-theory on arbitrary manifolds and topological insulators, Koushik Ray and Siddhartha Sen, arXiv:1408.4898 [cond-mat.str-el]

T. Eguchi, P. B. Gilkey and A. J. Hanson, Gravitation, Gauge Theories and Differential Geometry, Phys. Rep. **66**, 213 (1980) gives and introduction to topology, differential geometry and includes a discussion on K-theory.

Chapter 4

Boundary Conditions and
Self-Adjoint Extensions

We are familiar with the idea that in order to obtain solutions of differential equations, we need to impose boundary conditions and that the nature of the solutions depend on which boundary conditions we impose. For example, the spectrum of bosons and fermions in quantum mechanics obtained by solving Schroedinger's equation are different as the boundary conditions in these two cases are different. So the natural question that immediately arises is what are the possible choices of boundary conditions for a given operator representing an observable in quantum mechanics. In order to address this issue we must remember that observables in quantum mechanics should have real eigenvalues, which is necessary for a unitary time evolution. These operators are conventionally called Hermetian or self-adjoint, although, as we shall see below, these two concepts are not identical. In view of this, we can thus ask the question what are the possible boundary conditions that can be imposed on an operator in quantum mechanics such that it is self-adjoint? The answer to this question can be obtained from the pioneering work of von Neumann on self-adjoint extensions of operators in quantum mechanics. Self-adjoint extensions are known to play important roles in a variety of physical contexts including Aharonov-Bohm effect, two and three dimensional delta function potentials, anyons, anomalies, ζ-function renormalization, particle statistics in one dimension, black holes and integrable models. In this Chapter we shall first explain the basic ideas of self-adjoint extension and their relation to boundary conditions. We shall then explain the method of von Neumann through various examples. Finally we shall apply the method of self-adjoint extensions to some interesting physics problems.

4.1   Basic Ideas

We start with some basic definitions and results necessary for our purposes. We shall denote a generic operator by the symbol T. The operator T act on elements of a Hilbert space $\mathcal{H}$. As mentioned before, in order to obtain the spectrum of observables such as momentum or Hamiltonian in quantum mechanics, we need to impose suitable boundary conditions. The boundary conditions implies that T does not act on all of $\mathcal{H}$, but on a subset $D(T) \subset \mathcal{H}$ such that the elements of $D(T)$ obey the boundary conditions. Clearly $D(T)$ is not the full Hilbert space $\mathcal{H}$ as there will in general be many elements in $\mathcal{H}$ which would not obey the boundary conditions. The subset $D(T)$ is called the domain of T and it is an integral part of the definition of the operator denoted by T.

We now give a few definitions necessary for our purpose. Let the inner product of two elements $\alpha, \beta \in \mathcal{H}$ be denoted by (α, β). An operator T is called **symmetric** if the relation

$$(\phi, T\psi) = (T\phi, \psi)$$

holds for all elements $\phi, \psi \in D(T)$. Note that this relation is sometimes used to define self-adjoint operators in elementary quantum mechanics. Strictly speaking this condition only defines a symmetric operator and we shall see the difference between this relation and the condition for self-adjointness below.

Let T^* denote the operator adjoint to T. For our purpose, T^* and T have the same expression as differential operators. In order to specify T^* fully, we need to define its domain $D(T^*)$. For that, consider a symmetric operator T and let $\psi \in D(T)$. Now find *all* elements $\phi \in \mathcal{H}$ such that

$$(\phi, T\psi) = (T\phi, \psi).$$

The set of all such $\phi \in \mathcal{H}$ defines the domain of the adjoint $D(T^*)$. Note that $D(T^*)$ defined in this way is in general different from $D(T)$. Finally, an operator T is called **self-adjoint** if and only if

$$T = T^* \quad \text{and} \quad D(T) = D(T^*).$$

To make things clear we consider a very simple example, namely that of a the momentum operator P on a line interval. It is defined as

$$P = -i\frac{d}{dx}$$

where $x \in [0, 1]$. We define the domain of P as

$$D(P) = \{\psi(0) = \psi(1) = 0, \ \psi \text{ absolutely continuous and } \psi \in L^2[0,1]\},$$

where $L^2[0,1]$ denotes the set of square integrable functions on the line interval $[0,1]$. The domain $D(P)$ simply encodes the boundary conditions on the wavefunctions relevant for the momentum operator P.

We wish to ask the following questions :

(1) Is the operator P symmetric in the given domain $D(P)$?
(2) What is the domain of the adjoint $D(P^*)$?
(3) Is P self-adjoint, i.e. is $D(P) = D(P^*)$?

In order to answer these questions, consider the quantity

$$\Delta_P \equiv (\phi, P\psi) - (P\phi, \psi) = -i \left(\int_0^1 \phi^* \frac{d\psi}{dx} - \int_0^1 \frac{d\phi^*}{dx} \psi \right) dx$$
$$= -i[\phi^*(1)\psi(1) - \phi^*(0)\psi(0)],$$

where ϕ, ψ at the moment are general elements of the Hilbert space $\mathcal{H}$. In order to check if P is a symmetric operator in the domain $D(P)$, consider arbitrary $\phi, \psi \in D(P)$. The chosen boundary conditions imply that $\phi(1) = \phi(0) = 0$ and $\psi(1) = \psi(0) = 0$. Thus we have

$$\Delta_P = -i[\phi^*(1)\psi(1) - \phi^*(0)\psi(0))] = 0, \quad \forall \phi, \ \psi \in D(P),$$

which means P is symmetric in $D(P)$.

Lets us now address the second question, namely what is the domain $D(P^*)$ of the adjoint operator. For that, consider $\psi \in D(P)$ and ask what are all possible ϕ such that $\Delta_P = 0$. Since $\psi \in D(P), \psi(1) = \psi(0) = 0$. We thus see that $\Delta_P = 0$ is satisfied for any $\psi \in D(P)$ without any condition on ϕ. This means,

$$D(P^*) = \{\phi, \text{ absolutely continuous and } \psi \in L^2[0,1]\}$$

Finally we address the third question posed above. From the expressions of $D(P)$ and $D(P^*)$, it is easy to see from the above discussion that the two domains are not equal. This means that the momentum operator as defined above is not a self-adjoint operator. In order to further verify this statement, let us recall that if an operator is self-adjoint, then it must have a real spectrum. It is easy to verify that the eigenvalue equation

$$P\psi = \lambda\psi, \quad \psi(1) = \psi(0) = 0,$$

has no solutions for the real eigenvalue λ, which is in agreement with the above observation.

We have thus seen that the momentum operator P with the domain $D(P)$ is not a self-adjoint operator. The next question is whether there exists some other domain in which P is indeed self-adjoint? In order to address this question, let us consider a different domain for P given by

$$D_\theta(P) = \{\psi(1) = e^{i\theta}\psi(0), \ \psi \text{ absolutely continuous and } \psi \in L^2[0,1]\},$$

where $\theta \in R$ mod 2π. We shall also assume that $\psi(0) \neq 0$ since in that case, $D_\theta(P)$ reduces to $D(P)$. We again ask the previous three questions for the momentum operator in this domain $D_\theta(P)$.

To see if P is symmetric in $D_\theta(P)$, again consider arbitrary elements $\phi, \psi \in D_\theta(P)$. In this case,

$$\Delta_P = -i[\phi^*(1)\psi(1) - \phi^*(0)\psi(0)] = -i[e^{-i\theta}\phi^*(0)e^{i\theta}\psi(0) - \phi^*(0)\psi(0)] = 0$$

for all $\phi, \psi \in D_\theta(P)$. Thus P is symmetric in $\phi, \psi \in D_\theta(P)$.

Let us now find the domain of the adjoint in this case. As before, consider $\psi \in D_\theta(P)$, which means $\psi(1) = e^{i\theta}\psi(0)$. In order to find the domain of the adjoint we have to find all $\phi \in \mathcal{H}$ by solving the equation

$$\Delta_P = -i[\phi^*(1)\psi(1) - \phi^*(0)\psi(0)] = 0.$$

This implies that

$$\psi(0)\,[\phi^*(1)e^{i\theta} - \phi^*(0)] = 0$$
$$\implies \phi(1) = e^{i\theta}\phi(0) \quad \text{as} \ \ \psi(0) \neq 0$$

We therefore see that

$$D_\theta(P^*) = \{\phi(1) = e^{i\theta}\phi(0), \ \phi \text{ absolutely continuous and } \phi \in L^2[0,1]\},$$

that is, $D_\theta(P^*) = D_\theta(P)$. This also answers the third question as it shows that P is indeed self-adjoint in the domain $D_\theta(P)$.

Finally, we can check that the eigenvalue equation for P in this case given by

$$P\psi = \lambda\psi, \quad \psi(1) = e^{i\theta}\psi(0),$$

has the solutions

$$\psi_\theta(x) = e^{i\lambda x}, \quad \lambda_n = \theta + 2n\pi$$

where $n \in Z$. It may be noted that for each value of the parameter θ, the spectrum is different. We thus have a one parameter family of inequivalent quantizations of the momentum operator on a line interval.

What is the physical meaning of the self-adjoint extension? Recall that the boundary condition is $\psi(1) = e^{i\theta}\psi(0)$. This implies that the probability densities are equal at $x = 0$ and $x = 1$. We can thus identify the two points and compactify the line interval into a circle. The parameter θ can be interpreted as magnetic flux threading the circle. This gives a physical interpretation of the self-adjoint extension parameter for this particular problem.

4.2 von Neumann's Method of Self-Adjoint Extension

We have seen that the momentum operator on a line interval is not self-adjoint in a particular domain, but with a suitable choice of domain it can be made self-adjoint. The domain of self-adjointness of P was chosen in an ad hoc fashion, which in general is not always possible. Now we shall discuss a method due to von Neumann which for a given operator T with a domain $D(T)$ answers the following questions :

(1) Is T self-adjoint in $D(T)$?
(2) If not, can it be made self-adjoint?
(3) If it can be made self-adjoint, what is the corresponding domain?

In what follows, we shall not give any derivation or proofs of the technique due to von Neumann, but shall illustrate them with the example of the momentum operator.

We start with some definitions useful for our purpose. For a given symmetric operator T, consider the equations

$$T^*\phi_+ = +i\phi_+$$
$$T^*\phi_- = -i\phi_-$$

and let $n_\pm$ denote the number of linearly independent square integrable solutions with eigenvalues $\pm i$ respectively. The pair (n_+, n_-) are called the deficiency indices for the operator T. What is their significance? Before giving a formal answer to this question, note that if an operator is self-adjoint, then it is expected to have only real eigenvalues. Thus the existence of $\pm i$ in the spectrum is a measure of the deviation from self-adjointness. Hence it is plausible that finite deficiency indices of an operator would serve as a measure of its deviation from self-adjointness. This intuitive idea is indeed true and any operator T can be classified in terms of its deficiency indices as follows :

1) T is (essentially) self-adjoint iff $(n_+, n_-) = (0, 0)$.
2) T is not self-adjoint but has self-adjoint extensions iff $n_+ = n_-$.
3) If $n_+ \neq n_-$, then T has no self-adjoint extensions.

Before proceeding further, let us illustrate these concepts with our example of the momentum operator. Consider first the operator P with the domain $D(P)$. We have already found the corresponding $D(P^*)$. In order to find the deficiency indices, we have to solve the equations

$$-i\frac{d\phi_\pm}{dx} = \pm i\phi_\pm, \quad \phi_\pm(x) \in D(P^*).$$

These equations have square integrable solutions in the interval $[0, 1]$ given by

$$\phi_+(x) = C_+ e^{-x}, \quad \Longrightarrow \quad n_+ = 1$$
$$\phi_-(x) = C_- e^{x}, \quad \Longrightarrow \quad n_- = 1,$$

where $C_\pm$ are the normalization constants.. We therefore see that in this case, $n_+ = n_- = 1$, which means the P is not self-adjoint in $D(P)$ but has self-adjoint extensions, which agrees with what we found before.

What would be the deficiency indices when P is defined with the domain $D_\theta(P)$? In this case we have seen that $D_\theta(P^*) = D_\theta(P)$. Hence, the deficiency indices are obtained by solving the equations

$$-i\frac{d\phi_\pm}{dx} = \pm i\phi_\pm, \quad \phi_\pm(1) = e^{i\theta}\phi_\pm(0).$$

It is easy to see that these equations have no square integrable solutions for real values of the parameter θ, which implies that $n_+ = n_- = 0$. This implies that P is essentially self-adjoint in $D_\theta(P)$. This too agrees with what we have found before.

Let us now consider symmetric operator T in a domain $D(T)$ with deficiency indices $n_+ = n_- = n$ (say). The above discussion tells us that T is not self-adjoint but admits self-adjoint extensions. This means that we can find a suitable domain in which T is self-adjoint. How do we find such a domain? Here we shall again state the result due to von Neumann and illustrate it with our example. For such an operator T, von Neumann's theory tells us that the domain of self-adjointness is given by

$$D_U(T) = \{\phi + \phi_+ + U\phi_- | \phi \in D(T) \text{ and } U \text{ is a } n \text{ x } n \text{ unitary matrix}\}$$

We shall now illustrate this for the case of the momentum operator P. We start with the operator defined in the domain $D(P)$. We have seen that in this case, $n_+ = n_- = 1$ which means that it is not self-adjoint in $D(P)$. In the previous section we just stated what the domain of self-adjointness of P should be in an ad hoc fashion. What we shall do now is to derive that domain using the prescription of von Neumann.

In order to proceed, we write the properly normalized solutions $\phi_\pm$ as

$$\phi_+(x) = \frac{\sqrt{2}e}{\sqrt{e^2 - 1}}e^{-x},$$
$$\phi_-(x) = \frac{\sqrt{2}}{\sqrt{e^2 - 1}}e^{x}.$$

For the momentum operator defined in the domain $D(P)$, $n_\pm = 1$. Thus U in this case is a 1 x 1 unitary matrix, which can be written as a phase $e^{i\gamma}$, where $\gamma \in R$ mod 2π. Now if $\psi(x)$ is an arbitrary element of the domain $D_\gamma(P)$, then von Neumann's prescription tells us that,

$$\psi(x) = \phi(x) + \phi_+(x) + e^{i\gamma}\phi_-(x),$$

where $\phi(x) \in D(P)$, i.e. $\phi(0) = \phi(1) = 0$ and $\phi_\pm(x)$ are given above. Using this we find that

$$\psi(1) = \phi_+(1) + e^{i\gamma}\phi_-(1)$$
$$\psi(0) = \phi_+(0) + e^{i\gamma}\phi_-(0),$$

from which it follows that

$$\left|\frac{\psi(1)}{\psi(0)}\right|^2 = 1$$

or equivalently,

$$\psi(1) = e^{i\theta}\psi(0),$$

where $\theta \in R$ mod 2π. This is nothing but the condition which defined the domain $D_\theta(P)$, which was introduced earlier in an ad hoc fashion. We have thus obtained the correct domain of self-adjointness for the momentum operator on a line interval using the method of von Neumann.

Before moving on, consider the case of the momentum operator defined on the half-line R_+ with the boundary condition $\psi(0) = 0$ and $\psi \in L^2(R_+)$. In this case, $\phi_+(x) \sim e^{-x}$ is square integrable as $x \to \infty$, but $\phi_-(x) \sim e^x$ is not. Thus, for a momentum operator on a half line, $n_+ = 1$ while $n_- = 0$. Since $n_+ \neq n_-$, the momentum operator on the half line is neither self-adjoint nor can it be made self-adjoint by a suitable choice of domain. This is the reason why in quantum mechanics, momentum on half line is not a good operator. There is an interesting implication of this result. Imagine that we have a system whose energy E is bounded from below. This is the basic requirement of any well defined physical system. Without loss of generality, the minimum value of the energy can be taken to be zero. From the time-energy uncertainty relation, we can think that time could be an operator and it could have a representation given by $\hat{t} = -i\frac{d}{dE}$. Since $0 \leq E < \infty$, this situation is exactly similar to that of the momentum operator on the half line, and by the same logic, time can never be represented as a self-adjoint operator in quantum mechanics for physical systems whose energies are bounded from below.

This concludes our discussion of the self-adjointness of the momentum operator. Next we shall the simplest possible Hamiltonian of a free particle on a half line.

4.2.1   *Free Particle on a Half-Line*

Consider the Hamiltonian for a free particle of unit mass on a half-line $x \geq 0$, $x \in R^+$, given by

$$H_0 = -\frac{d^2}{dx^2}.$$

Consider the subset A of the Hilbert space $\mathcal{H}$ given by

$$A = \{\phi \mid \phi \in L^2(R^+), H\phi \in L^2(R^+), \ \phi' \text{ absolutely continuous}\}.$$

We define the domain of the operator H_0 as

$$D(H_0) = \{\phi \mid \phi \in A, \ \phi(0) = \phi'(0) = 0\}.$$

As before, we ask the following questions :

(1) Is H_0 symmetric in $D(H_0)$?
(2) What is $D(H_0^*)$?
(3) Is H_0 self-adjoint in $D(H_0)$?

To address these questions, lets us consider the expression

$$\Delta_{H_0} \equiv (\phi, H_0\psi) - (H_0\phi, \psi) = -\int_0^\infty \left(\phi^* \frac{d^2\psi}{dx^2} - \frac{d^2\phi^*}{dx^2}\psi\right) dx$$

$$= \phi^*(0)\frac{d\psi(0)}{dx} - \frac{d\phi^*(0)}{dx}\psi(0),$$

where ϕ, ψ at the moment are general elements of the Hilbert space $\mathcal{H}$ which go to zero at infinity so that they are square integrable. In order to check if H_0 symmetric in $D(H_0)$, consider arbitrary $\phi, \psi \in D(H_0)$. In that case, we know that $\phi(0) = \phi'(0) = \psi(0) = \psi'(0) = 0$. Thus we have

$$\Delta_{H_0} = \phi^*(0)\frac{d\psi(0)}{dx} - \frac{d\phi^*(0)}{dx}\psi(0) = 0 \ \ \forall \phi, \psi \in D(H_0),$$

which means H_0 is symmetric in $D(H_0)$.

Let us now address the second question, namely what is the domain $D(H_0^*)$ of the adjoint operator H_0^*. For that, consider $\psi \in D(H_0)$ and ask what are all possible ϕ such that $\Delta_{H_0} = 0$. Since $\psi \in D(H_0), \psi(0) = \psi'(0) = 0$. We thus see that $\Delta_{H_0} = 0$ is satisfied for any $\psi \in D(H_0)$ without any condition on ϕ. This means that

$$D(H_0^*) = \{\phi \mid \phi \in A\},$$

without any further condition on ϕ.

Finally, we see that just as in the case for the momentum operator on half-line, here also $D(H_0^*) \neq D(H_0)$ as subsets of $\mathcal{H}$. This implies that H_0 is not self-adjoint in $D(H_0)$.

It is now clear that the reason for the above is that the conditions defining $D(H_0)$ are too restrictive. We can therefore expect that if the defining conditions on $D(H_0)$ are made less restrictive, i.e. if $D(H_0)$ is enlarged, the operator H_0 may become self-adjoint. With that in mind, consider the following domain

$$D_\gamma(H_0) = \{\phi \mid \phi \in A, \phi'(0) = \gamma\phi(0)\},$$

where $\gamma \in R$. With this new definition of the domain, it now easy to check that

(1) H_0 is symmetric in $D_\gamma(H_0)$.
(2) $D_\gamma(H_0^*) = D_\gamma(H_0)$.

Thus H_0 is self-adjoint in $D_\gamma(H_0)$. This shows that enlarging the domain of H_0 has reduced the domain of H_0^* such that now they are equal and the operator is self-adjoint.

We shall now proceed with the method of von Neumann to analyze the operator H_0 defined with the domain $D(H_0)$. In order to find the deficiency indices, we have to find square integrable solutions of the equations

$$-\frac{d^2\phi_\pm}{dx^2} = \pm i\phi_\pm, \quad \phi_\pm \in D(H_0^*).$$

The solutions are

$$\phi_+ = 2^{\frac{1}{4}}\exp\left(\frac{(i-1)}{\sqrt{2}}x\right), \quad \Longrightarrow \quad n_+ = 1$$

$$\phi_- = 2^{\frac{1}{4}}\exp\left(\frac{-(i+1)}{\sqrt{2}}x\right), \quad \Longrightarrow \quad n_- = 1$$

This shows that the Hamiltonian of a free particle on a half-line defined in the domain $D(H_0)$ have deficiency indices $(n_+, n_-) = (1, 1)$. According to the previous discussion, this Hamiltonian is thus not self-adjoint in the domain $D(H_0)$ but admits a one-parameter family of self-adjoint extensions. Following the prescription of von Neumann, in this case also we can find the domain of self-adjointness of H_0. We expect that H_0 would be self-adjoint in the domain

$$D_\beta(H_0) = \{\psi = \phi + \phi_+ + \beta\phi_- \mid \phi \in D_((H_0)\},$$

where $\phi_\pm$ are given above respectively and β is a unitary 1 x 1 matrix, or a pure phase. Using a similar analysis as given for the momentum operator, it can be shown here as well that if $\psi \in D_\beta(H_0)$, then

$$\psi'(0) = \gamma\psi(0),$$

where $\gamma \in R$ and γ is a function of the phase β. This condition is the same as that on the wavefunction belonging to $D_\gamma(H_0)$. Recall the $D_\gamma(H_0)$ was defined in an ad hoc fashion before. The method of von Neumann confirms that $D_\gamma(H_0)$ is the correct domain of self-adjointness.

Let us find find the solutions of the Schrodinger's equation

$$-\frac{d^2\psi}{dx^2} = E\psi, \quad \psi'(0) = \gamma\psi(0).$$

First consider the case when $E \equiv p^2 > 0$. In this case the general solution is given by

$$\psi = Ce^{-ipx} + De^{ipx},$$

where C, D are constants. Imposing the boundary condition gives

$$\frac{C}{D} = \frac{ip + \gamma}{ip - \gamma}$$

which leads to the solution

$$\psi = C\left(e^{-ipx} + \frac{ip + \gamma}{ip - \gamma}e^{ipx}\right).$$

If $R \equiv \frac{ip+\gamma}{ip-\gamma}$, we have $RR^* = 1$, i.e. $R = e^{i\theta(\gamma)}$ where $\theta(\gamma)$ is the phase of the reflected wave that depends on the parameter γ. This solutions represents the scattering states of the problem.

Consider next the case where $E \equiv -\mathcal{E}$ and ask the question if the system admits square integrable solutions, or bound states. We try the ansatz

$$\psi = De^{-\eta x}, \quad \eta \in R, \quad \eta > 0,$$

where D is a constant and η has been chosen positive for square integrability. With $E = -\mathcal{E}$ we have,

$$\mathcal{E} = \eta^2.$$

Imposing the boundary condition we get,

$$\eta = -\gamma.$$

Since $\eta > 0$, we must have $\gamma < 0$ for square integrable bound states to exist. Thus the complete solution of the bound state eigenvalue problem is given by

$$\psi(x) = \sqrt{-2\gamma}e^{\gamma x}, \quad \gamma < 0, \quad E = -\gamma^2.$$

This result is indeed striking. How is it possible for a free particle to admit a bound state. A bound state necessarily requires a scale, which a free particle Hamiltonian does not contain. The answer to this apparent puzzle lies in the fact that the boundary condition which defines the domain of self-adjointness supplies the necessary scale. To see this more clearly, note that α has dimensions of inverse length and is the relevant dimensionful parameter related to the bound state energy. We can now go back and ask how such a dimensionful parameter appears in the prescription of von Neumann. To understand this, look at the equations from which we obtained the deficiency indices for the free particle Hamiltonian. For consistency, the right hand side of those equations must have a constant with dimension length squared, whose value has conveniently been set equal to unity. Nevertheless, the dimensionful constant actually enters the definition of the domain or equivalently the boundary conditions. Such a dimensionful constant breaks the scale invariance of the problem and produces the bound state for when $\alpha < 0$. This is perhaps the simplest example of quantum mechanical breakdown of scale invariance, or scaling anomaly. This concludes our discussion on a free particle on the half-line.

4.2.2 *Inverse Square Interaction*

So far we have considered Hamiltonians without interaction. Now we will discuss a system with an inverse square interaction. Such an interaction occurs in a variety of physical systems, some of which will be considered later in detail. For the moment we shall focus our attention on the general properties of the system.

We start with the Hamiltonian given by

$$H = \left[-\frac{\partial^2}{\partial r^2} + \frac{\alpha}{r^2} \right],$$

where $r \in (0, \infty) \equiv R^+$, which can be identified with the radial coordinate. The parameter α is a real constant. Define

$$A = -\frac{d}{dr} - \frac{\gamma}{r},$$

where γ is a real constant. The formal adjoint of A is given by

$$A^\dagger = \frac{d}{dr} - \frac{\gamma}{r}.$$

We want to find the condition under which we can write

$$H = AA^\dagger = \left(-\frac{d}{dr} - \frac{\gamma}{r}\right)\left(\frac{d}{dr} - \frac{\gamma}{r}\right),$$

This decomposition has the formal structure of a product of an operator and its adjoint and therefore if H can be written in this form, then it will be formally a positive operator. Comparing the two expressions for the Hamiltonian, we find

$$\alpha = \gamma(\gamma - 1)$$

Hence the minimum value of α for which the operator H can be written as $A^\dagger A$ is given by $\alpha = -\frac{1}{4}$, which corresponds to $\gamma = \frac{1}{2}$. Therefore, if $\alpha < -\frac{1}{4}$, then the operator H is not even a formally positive quantity. Thus $\alpha = -\frac{1}{4}$ corresponds to a critical value of the inverse square coupling.

We shall first discuss the case where $\alpha \geq -\frac{1}{4}$. To that end we write $\alpha = \nu^2 - \frac{1}{4}$ with $\nu \geq 0$. The Hamiltonian then takes the form

$$H = \left[-\frac{\partial^2}{\partial r^2} + \frac{\nu^2 - \frac{1}{4}}{r^2} \right].$$

Our goal is to solve the Schrodinger's equation

$$H\psi = E\psi.$$

This Hamiltonian is an unbounded differential operator defined in R^+. It is a symmetric operator on the domain $D(H) \equiv \{\phi(0) = \phi'(0) = 0, \phi, \phi'$ absolutely continuous, $\phi \in L^2(dr)\}$. We would next like to determine if H is self-adjoint, for which we treat the cases $\nu \neq 0$ and $\nu = 0$ separately.

$\nu \neq 0$

As discussed before, the deficiency indices $n_\pm$ are determined by the number of square-integrable solutions of the equations

$$H^*\phi_\pm = \pm i\phi_\pm,$$

respectively, where H^* is the adjoint of H. Note that H^* is given by the same differential operator as H. From dimensional considerations we see that the right hand side of the equation defining the deficiency indices should be multiplied with a constant with dimension of length^{-2}. We shall henceforth choose the magnitude of this constant to be unity by appropriate choice of units. The physical relevance of this constant will be discussed later.

The solutions of the equation for the deficiency indices are given by

$$\phi_+(r) = r^{\frac{1}{2}} H_\nu^{(1)}(re^{i\frac{\pi}{4}}),$$
$$\phi_-(r) = r^{\frac{1}{2}} H_\nu^{(2)}(re^{-i\frac{\pi}{4}}),$$

where H_ν's are Hankel functions. The functions $\phi_\pm$ are bounded as $r \to \infty$. When $r \to 0$, they behave as

$$\phi_+(r) \to \frac{i}{\sin\nu\pi} \left[\frac{r^{\nu+\frac{1}{2}}}{2^\nu} \frac{e^{-i\frac{3\nu\pi}{4}}}{\Gamma(1+\nu)} - \frac{r^{-\nu+\frac{1}{2}}}{2^{-\nu}} \frac{e^{-i\frac{\nu\pi}{4}}}{\Gamma(1-\nu)} \right],$$

$$\phi_-(r) \to \frac{i}{\sin\nu\pi} \left[-\frac{r^{\nu+\frac{1}{2}}}{2^\nu} \frac{e^{i\frac{3\nu\pi}{4}}}{\Gamma(1+\nu)} + \frac{r^{-\nu+\frac{1}{2}}}{2^{-\nu}} \frac{e^{i\frac{\nu\pi}{4}}}{\Gamma(1-\nu)} \right].$$

We see that $\phi_\pm$ are not square integrable when $\nu^2 \geq 1$. In this case H has deficiency indices $(0,0)$ and is (essentially) self-adjoint on the domain $D(H)$. On the other hand, both $\phi_\pm$ are square integrable when either $-1 < \nu < 0$ or $0 < \nu < 1$. We therefore see that for any value of ν in these ranges, H has deficiency indices $(1,1)$. In this case, H is not self-adjoint on the domain $D(H)$ but admits self-adjoint extensions. The deficiency subspaces $K_\pm$ in this case are one dimensional and are spanned by the functions $\phi_\pm$. The unitary maps from K_+ into K_- are parameterized by e^{iz} where $z \in R \pmod{2\pi}$. The operator H is self-adjoint in the domain $D_z(H) = D(H) \oplus a\{\phi_+(r) + e^{iz}\phi_-(r)\}$ where a is an arbitrary complex number.

We would now like to obtain the spectrum of H in the domain $D_z(H)$. We start with the discussion of the scattering states given by the positive energy solutions of Hamiltonian H. For that we set $E = q^2$, where q is a real positive parameter. The general solution of the Schrodinger's equation can be written as

$$\psi(r) = r^{\frac{1}{2}} \left[a(q)J_\nu(qr) - b(q)J_{-\nu}(qr) \right],$$

where $a(q)$ and $b(q)$ are two as yet undetermined coefficients. J_ν refers to the Bessel function of order ν. Note that in the limit $r \to 0$,

$$\phi_+(r) + e^{iz}\phi_-(r) \to \frac{i}{\sin\nu\pi} \left[\frac{r^{\nu+\frac{1}{2}}}{2^\nu} \frac{(e^{-i\frac{3\nu\pi}{4}} - e^{i(z+\frac{3\nu\pi}{4})})}{\Gamma(1+\nu)} + \frac{r^{-\nu+\frac{1}{2}}}{2^{-\nu}} \frac{(e^{i(z+\frac{\nu\pi}{4})} - e^{-i\frac{\nu\pi}{4}})}{\Gamma(1-\nu)} \right]$$

and

$$\chi(r) \to \frac{r^{\nu+\frac{1}{2}}}{2^\nu} \frac{a(q)q^\nu}{\Gamma(1+\nu)} - \frac{r^{-\nu+\frac{1}{2}}}{2^{-\nu}} \frac{b(q)q^{-\nu}}{\Gamma(1-\nu)}.$$

If $\psi(r) \in D_z(H)$, then the coefficients of of $r^{\nu+\frac{1}{2}}$ and $r^{-\nu+\frac{1}{2}}$ in the above equations must match. Comparing these coefficients we get

$$\frac{a(q)}{b(q)} = \frac{\sin(\frac{z}{2} + 3\pi\frac{\nu}{4})}{\sin(\frac{z}{2} + \pi\frac{\nu}{4})} q^{-2\nu}.$$

Next we calculate the S-matrix and the associated phase shift. In the limit $r \to \infty$, the leading term in the asymptotic expansion of $\chi(r)$ is given by

$$\psi(r) \to \frac{1}{\sqrt{2\pi q}} e^{iqr} \left[a(q)e^{-i(\nu+\frac{1}{2})\frac{\pi}{2}} - b(q)e^{i(\nu-\frac{1}{2})\frac{\pi}{2}} \right]$$

$$+ \frac{1}{\sqrt{2\pi q}} e^{-iqr} \left[a(q)e^{i(\nu+\frac{1}{2})\frac{\pi}{2}} - b(q)e^{-i(\nu-\frac{1}{2})\frac{\pi}{2}} \right].$$

By dividing the coefficient of outgoing wave (e^{iqr}) by that of the incoming wave (e^{-iqr}), we obtain the the S-matrix and phase shift $\delta(q)$ as

$$S(q) \equiv e^{2i\delta(q)} = \frac{a(q)e^{-i(\nu+\frac{1}{2})\frac{\pi}{2}} - b(q)e^{i(\nu-\frac{1}{2})\frac{\pi}{2}}}{a(q)e^{i(\nu+\frac{1}{2})\frac{\pi}{2}} - b(q)e^{-i(\nu-\frac{1}{2})\frac{\pi}{2}}}.$$

Finally, we get

$$S(q) = e^{2i\delta(q)} = \frac{q^{-\nu}\sin(\frac{z}{2}+3\pi\frac{\nu}{4})e^{-i(\nu+\frac{1}{2})\frac{\pi}{2}} - q^{\nu}\sin(\frac{z}{2}+\pi\frac{\nu}{4})e^{i(\nu-\frac{1}{2})\frac{\pi}{2}}}{q^{-\nu}\sin(\frac{z}{2}+3\pi\frac{\nu}{4})e^{i(\nu+\frac{1}{2})\frac{\pi}{2}} - q^{\nu}\sin(\frac{z}{2}+\pi\frac{\nu}{4})e^{-i(\nu-\frac{1}{2})\frac{\pi}{2}}}.$$

Let us now consider the bound state solutions. For any given value of ν in the allowed range the S-matrix has a pole on the positive imaginary axis of the complex q-plane. Such a pole indicates the existence of a bound state for the effective Hamiltonian H. By taking $q = i\rho$ as the pole for the S-matrix, one can easily derive the corresponding bound state energy $E = -\rho^2$ as

$$E = -\left[\frac{\sin(\frac{z}{2}+3\pi\frac{\nu}{4})}{\sin(\frac{z}{2}+\pi\frac{\nu}{4})}\right]^{\frac{1}{\nu}}.$$

Thus we see that for a given value of ν within the allowed range, H admits a single bound state with energy given as above. It may be noted that for a fixed ν, the bound state exists only for those values of z such that the quantity $\frac{\sin(\frac{z}{2}+3\pi\frac{\nu}{4})}{\sin(\frac{z}{2}+\pi\frac{\nu}{4})}$ is positive. The corresponding bound state eigenfunction is given by

$$\chi(r) = Br^{\frac{1}{2}}H_\nu^{(1)}(i\sqrt{|E|}r),$$

where B is the normalization constant.

Note that when $\nu^2 \geq 1$, H is essentially self-adjoint in the domain $D(H)$. In this case, the scattering state solutions are given by

$$\chi(r) = a(q)r^{\frac{1}{2}}J_\nu(qr).$$

No bound states exist in this case.

3.2 $\nu = 0$

In this case, the solutions of the equations

$$H^*\psi_\pm = \pm i\psi_\pm$$

are given by

$$\psi_+(r) = r^{\frac{1}{2}}H_0^{(1)}(re^{i\frac{\pi}{4}}),$$
$$\psi_-(r) = r^{\frac{1}{2}}H_0^{(2)}(re^{-i\frac{\pi}{4}})$$

respectively. $\psi_\pm$ are bounded functions as $r \to \infty$. In order to find their behaviour for small r, we first note that in the limit $r \to 0$,

$$J_0(r) = 1 + \mathcal{O}\left(r^2\right)$$

$$N_0(r) = \frac{2}{\pi}\left[\gamma - \ln 2 + \ln r\right] + \mathcal{O}\left(r^2 \ln r\right)$$

where γ is Euler's constant and N_0 is the Neumann function. We see that when $r \to 0$, $\psi_+(r)$ behaves as

$$\psi_+(r) \to \frac{2i}{\pi} r^{\frac{1}{2}} \ln r + r^{\frac{1}{2}}\left[\frac{1}{2} + \frac{2i}{\pi}(\gamma - \ln 2)\right].$$

In the same limit, $\psi_-(r)$ behaves as

$$\psi_-(r) \to -\frac{2i}{\pi} r^{\frac{1}{2}} \ln r + r^{\frac{1}{2}}\left[\frac{1}{2} - \frac{2i}{\pi}(\gamma - \ln 2)\right].$$

Thus we see that both $\psi_\pm(r)$ are square integrable functions. H therefore has deficiency indices $(1,1)$ and the corresponding deficiency subspaces $K_\pm$ are again 1-dimensional, spanned by the functions $\psi_\pm(r)$. As before, the operator H is not self-adjoint on $D(H)$ but admits a one-parameter family of self-adjoint extensions labeled by e^{iz} where $z \in R$ mod 2π. It is self-adjoint in the domain $D_z(H)$ which contains all the elements of $D(H)$ together with elements of the form $a(\psi_+(r) + e^{iz}\psi_-(r))$ where a is an arbitrary complex number.

The scattering states associated with positive energy solutions are given by

$$\psi(r) = r^{\frac{1}{2}} \left[a(q)J_0(qr) - b(q)N_0(qr)\right],$$

where $E = q^2$ and $a(q)$, $b(q)$ are two as yet undetermined coefficients. In order to find the ratio of these two coefficients, as before we use the fact that if H has to be self-adjoint, the eigenfunction $\psi(r)$ must belong to the domain $D_z(H)$. In the limit when $r \to 0$, we have

$$\psi(r) \to -\frac{2b(q)}{\pi} r^{\frac{1}{2}} \ln r + r^{\frac{1}{2}}\left[a(q) - \frac{2b(q)}{\pi} \ln q + \frac{2b(q)}{\pi}(\ln 2 - \gamma)\right].$$

Comparing the coefficients of $r^{\frac{1}{2}} \ln r$ we find

$$b(q) = -2\sin\frac{z}{2}.$$

Comparing the coefficients of $r^{\frac{1}{2}}$ we obtain

$$a(q) - \frac{2b(q)}{\pi} \ln q = \cos\frac{z}{2}.$$

Assuming that $z \neq 0$, we finally obtain

$$\frac{a(q)}{b(q)} = \frac{2}{\pi} \ln q - \frac{1}{2} \cot \frac{z}{2}.$$

Thus we find that for any generic value of the self-adjoint parameter $z \neq 0$, the ratio of $a(q)$ and $b(q)$ depends on the momentum q.

For calculating the S-matrix and phase shift for the above mentioned scattering process, we consider the leading term in the asymptotic expansion of $\psi(r)$ at $r \to \infty$ limit. This gives

$$\psi(r) \to \frac{1}{\sqrt{2\pi q}} e^{iqr} e^{-i\frac{\pi}{4}} \left[a(q) + ib(q) \right] + \frac{1}{\sqrt{2\pi q}} e^{-iqr} e^{i\frac{\pi}{4}} \left[a(q) - ib(q) \right].$$

By dividing the coefficient of outgoing wave (e^{iqr}) by that of the incoming wave (e^{-iqr}), we obtain the the S-matrix and phase shift as

$$S(q) \equiv e^{2i\delta(q)} = e^{-i\frac{\pi}{2}} \frac{\frac{2}{\pi} \ln q - \frac{1}{2} \cot \frac{z}{2} + i}{\frac{2}{\pi} \ln q - \frac{1}{2} \cot \frac{z}{2} - i}.$$

When $z \neq 0$, the S-matrix again has a single pole on the positive imaginary axis of the complex q-plane. Such a pole naturally indicates the existence of a bound states for the effective Hamiltonian $\tilde{H}$. By taking $q = i\rho$ as the pole for the S-matrix, we obtain the corresponding bound state energy $E = -\rho^2$ as

$$E = -\exp \left[\frac{\pi}{2} \cot \frac{z}{2} \right].$$

The corresponding eigenfunction is given by

$$\psi(r) = C r^{\frac{1}{2}} K_0 \left(\sqrt{|E|} r \right) = C \frac{i\pi}{2} r^{\frac{1}{2}} H_0^{(1)} \left(i\sqrt{|E|} r \right),$$

where C is a constant and K_0 is the modified Bessel function.

We have seen that for a given value of ν, each value of the parameter $z \pmod{2\pi}$ produces a different spectrum of H. The spectrum of H, and correspondingly the parameter space of the quantum theory is thus characterized by the pair (ν, z). Let us assume for the moment that $\nu \neq 0$. Consider now two pairs of parameters (ν, z) and (ν', z') corresponding to two different quantum theories. It is easily seen that the spectrum of these two theories are identical if $\nu' = -\nu$ and $z' = z + 2\pi\nu$. We therefore have an equivalence relation $(\nu, z) \sim (-\nu, z + 2\pi\nu)$ on the parameter space of H. The transformation $(\nu, z) \to (\nu', z')$ relating two different quantum theories is analogous to the duality symmetry in this system. In the case when $\nu = 0$, we automatically have $(\nu, z) = (\nu', z')$ for all values of z. We can thus say that the pair $(\nu = 0, z)$ defines the self-dual points in the parameter space.

As mentioned before, the classical system that we started with possesses scaling symmetry. However, in the presence of the self-adjoint extensions, the system admits bound state(s) and the phase shifts in the scattering sector depend explicitly on the momentum. These are indicative of the breakdown of the scaling symmetry at the quantum level. We first analyze this issue when $-1 < \nu \neq 0 < 1$. Let us consider the action of the scaling operator $\Lambda = \frac{-i}{2}(r\frac{d}{dr} + \frac{d}{dr}r)$ on an element $\phi(r) = \phi_+(r) + e^{iz}\phi_-(r) \in D_z(H)$. In the limit $r \to 0$, we have

$$\Lambda\phi(r) \to \frac{1}{\sin\nu\pi}\left[(1+\nu)\frac{r^{\nu+\frac{1}{2}}}{2^\nu}\frac{(e^{-i\frac{3\nu\pi}{4}} - e^{i(z+\frac{3\nu\pi}{4})})}{\Gamma(1+\nu)}\right.$$
$$\left. +(1-\nu)\frac{r^{-\nu+\frac{1}{2}}}{2^{-\nu}}\frac{(e^{i(z+\frac{\nu\pi}{4})} - e^{-i\frac{\nu\pi}{4}})}{\Gamma(1-\nu)}\right].$$

In order for $\Lambda\phi(r) \in D_z(H)$, we must have $\Lambda\phi(r) \sim C\phi(r)$ where C is a constant. However, the two terms on the right hand side of the above equation are multiplied by two different factors, namely by $(1 + \nu)$ and $(1 - \nu)$. Due to the presence of these different multiplying factors, we see that $\Lambda\phi(r)$ in general does not belong to $D_z(H)$. Scale invariance is thus broken at the quantum level for generic values of z. However for special choice of $z = -\frac{\nu\pi}{2}$, $\Lambda\phi(r) \in D_z(H)$ and the scaling symmetry is recovered. In addition, we find that the scaling symmetry is also preserved at the quantum level when $z = -\frac{3\nu\pi}{2}$. For these choices of z, the bound states do not exist and the S matrix becomes independent of the momentum. The scaling symmetry thus is anomalously broken due to the quantization for generic values of z. In the case when $\nu = 0$, a similar analysis as above again shows that the self-adjoint extension generically breaks the scaling symmetry. For this case, the scale invariance can be recovered at the quantum level only for $z = 0$.

4.2.3 *Inverse Square Potential at Strong Coupling*

We shall now consider an inverse square interaction where the coupling $\alpha < -\frac{1}{4}$. As discussed before, in this case, the Hamiltonian is no longer a formally positive operator. Consider now the eigenvalue problem

$$H\psi = E\psi.$$

Assume that H admits a bound state with energy $E = -E_0$. We will argue that E_0 cannot be finite. To see this, first note that the constant α is dimensionless and hence the operator H is scale invariant. Imagine now that there exists a wavefunction ψ_0 for which we have

$$H\psi_0 = -\left[\frac{d^2}{dr^2} + \frac{\alpha}{r^2}\right]\psi_0 = -E_0\psi_0.$$

Since α is dimensionless, we can scale r such that $r \to br$, where b is a positive constant. Under this scaling, we would have

$$H\psi_0 = -b^2 E_0\psi_0.$$

Thus by taking the constant $b \to \infty$, the energy of the system can be made arbitrarily negative. This means that the system is unbounded from below, or in other words unphysical. This is not surprising. The operator H has no dimensionful quantity. For such a system, there cannot be a bound state because there is no dimensionfull parameter in the system in terms of which the bound state energy can be calculated. This qualitative argument shows that a truly scale invariant system cannot admit a stable negative energy bound state.

How do we proceed in such a situation? It is clear that the instability is coming due to the singular nature of the interaction at short distance. As a first step then we introduce a short distance cutoff at $r = a$. We also write $\alpha = \mu^2 + \frac{1}{4}$ so that $-\alpha < -\frac{1}{4}$. In addition, we impose the boundary condition that

$$\psi(r = a) = 0.$$

Let us now consider the eigenvalue problem of the Hamiltonian H in the bound state sector of the theory. This is given by

$$H\psi = -\kappa^2\psi, \qquad \kappa \in R.$$

This can be written as

$$\frac{d^2\psi}{dz^2} - \left[\frac{\mu^2 + \frac{1}{4}}{r^2} - 1\right]\psi = 0, \qquad z = \kappa r.$$

The above equation is solved by

$$\psi(z) = \sqrt{z}K_{i\mu}(z),$$

where $K_{i\mu}(z)$ is the modified Bessel function of the third kind with order $i\mu$, which is purely imaginary. For small values of z and μ, the zeroes of this function, denoted by z_n are given by

$$z_n = e^{\left(\frac{-n\pi}{\mu}\right)} (2e^{-\gamma}) [1 + O(\mu)],$$

where γ is the Euler's constant and $n = 1, 2, ..., \infty$. Therefore the energy levels are given by

$$E_n = -e^{\left(\frac{-2n\pi}{\mu}\right)} \left[\frac{2}{ae^\gamma}\right]^2 [1 + O(\mu)].$$

We can immediately see from the above equation that if the cutoff a is taken to zero keeping μ fixed, the energy levels go to negative infinity.

In order to handle this divergence, we use the renormalization group approach. First we regard μ as a function of the cutoff a and ask how $\mu(a)$ must depend on a such that when the latter is taken to zero, the ground state energy E_1 remains independent of a. This is going to fix the dependence of $\mu(a)$ on a and hence give the β function. Let us define the β function as

$$\beta(\mu) = -a\frac{d\mu}{da}.$$

The condition that the ground state energy remains independent of the cutoff as the latter is taken to zero can be written as

$$a\frac{dE_1}{da} = 0$$

This gives

$$\beta(\mu) = -\frac{\mu^2}{\pi} + ...$$

Thus we see that our system has an ultraviolet stable fixed point at $\mu = 0$ which is same as $\alpha = -\frac{1}{4}$. This means that in the continuum limit, the renormalization group flow would drive the coupling to its critical value.

4.2.4 Application to Polar Molecules

As a physics example of the method of self-adjoint extensions, we now discuss the problem of electron capture by polar molecules. It has been experimentally observed that there are some discrepancies between what is expected and what is observed in electron scattering by polar molecules. Such an anomalous scattering is often attributed to the electron capture by the dipole field of the polar molecules. Such a bound state is normally expected if the dipole interaction is sufficiently strong, or the dipole moment of the molecule is above a certain critical value D_0, whose value has been calculated to be $D_0 = 1.63 \times 10^{-18}$ esu cm. This is consistent with most of the experimental data.

However, certain polar molecules such as H_2S and HCl have dipole moments less than D_0 and yet exhibit anomalous electron scattering. Hence

the theory which predicts the anomalous scattering when the dipole moment exceeds the critical value is not sufficient to explain these cases. In
order to offer a plausible explanation, recall first that the dipole interaction is primarily a long range effect. In a molecule, there also could be
various short range interactions which could contribute to the electron capture. Within a simple analytical framework, it is difficult to incorporate
all such interactions separately in the Hamiltonian. Their collective effect
may however be incorporated through appropriate boundary conditions.
The molecular interactions are such that the Hamiltonian is expected to
provide a unitary time evolution. Thus the problem of finding the boundary conditions reduces to that of finding all possible self-adjoint extensions
of the Hamiltonian. The analysis presented here is based on operator theory discussed in the previous sections. Using the method of von Neumann,
we shall show that a polar molecule can capture an electron even for certain ranges of the dipole moment below its critical value. In such cases, a
single bound state will be found, which is in qualitative agreement with the
experimental data.

Consider an electron of charge e and mass μ moving in a point dipole
field with dipole moment D. We take the z axis along the dipole moment.
Schrodinger's equation for the electron in the spherical polar coordinates
(r, θ, ϕ) can be written as

$$\left[-\frac{\hbar^2}{2\mu} \nabla^2 + \frac{eD}{r^2} \cos\theta \right] \Psi = E\Psi ,$$

where E is the energy. Consider the wavefunction $\Psi(r, \theta, \phi) = \frac{1}{r} R(r) \Theta(\theta) e^{im\phi}$, which satisfies the radial equation

$$H_r R(r) \equiv \left[-\frac{d^2}{dr^2} + \frac{\lambda}{r^2} \right] R(r) = \epsilon R(r) ,$$

where H_r is the radial Hamiltonian, $\epsilon = \frac{2\mu E}{\hbar^2}$ is the eigenvalue of H_r. For λ
we shall use the lowest eigenvalue of the angular equation, which is given
by

$$\lambda = -\frac{1}{6} d^2 + \frac{11}{1080} d^4 - \frac{133}{97200} d^6 + \dots ,$$

with $d = \frac{2\mu e D}{\hbar^2}$. It is generally assumed that in order for the operator H_r to
admit a bound state, the coefficient of the inverse square interaction must
be sufficiently negative, so that the condition given by $\lambda < -\frac{1}{4}$ is satisfied.
This assumption leads to the critical dipole moment $D \geq D_0 = 1.63 \times 10^{-18}$
esu cm, which corresponds to $\lambda = -\frac{1}{4}$. Using von Neumann's approach,

and with a very general assumption that the Hamiltonian is self-adjoint, we shall now show that it is possible to form bound states in this system with a weaker condition on λ and correspondingly for smaller values of D_0.

Note first that for any non-zero value of the dipole moment, the parameter $\lambda < 0$. We restrict our analysis to the range $-\frac{1}{4} \leq \lambda < 0$, where, as discussed before, the self-adjoint extension plays a role. The Hamiltonian H_r is a real symmetric operator on the domain $D(H_r) \equiv \{\phi(0) = \phi'(0) = 0, \phi, \phi'$ absolutely continuous$\}$. Following von Neumann's method, we now look for square integrable solutions of the equations equations

$$H_r^* \phi_\pm = \pm i \phi_\pm,$$

where H_r^* is the adjoint of H_r. In terms of the variable $\nu = \sqrt{\lambda + \frac{1}{4}}$, the solutions of the equations which gives the deficiency indices are given by

$$\phi_+(r) = r^{\frac{1}{2}} H_\nu^{(1)}(re^{i\frac{\pi}{4}}),$$
$$\phi_-(r) = r^{\frac{1}{2}} H_\nu^{(2)}(re^{-i\frac{\pi}{4}}),$$

where H_ν's are Hankel functions. These solutions have been chosen so that they are square integrable at infinity. For the moment we consider the case $\lambda \neq -\frac{1}{4}$. The functions $\phi_\pm$ are bounded as $r \to \infty$. When $r \to 0$, they behave as

$$\phi_+(r) \to C_1(\nu)r^{\nu+\frac{1}{2}} + C_2(\nu)r^{-\nu+\frac{1}{2}},$$
$$\phi_-(r) \to C_1^*(\nu)r^{\nu+\frac{1}{2}} + C_2^*(\nu)r^{-\nu+\frac{1}{2}},$$

where $C_1(\nu) = \frac{i}{\sin\nu\pi}\frac{1}{2^\nu}\frac{e^{-i\frac{3\nu\pi}{4}}}{\Gamma(1+\nu)}$, $C_2(\nu) = -\frac{i}{\sin\nu\pi}\frac{1}{2^{-\nu}}\frac{e^{-i\frac{\nu\pi}{4}}}{\Gamma(1-\nu)}$ and $C_1^*(\nu)$ and $C_2^*(\nu)$ are complex conjugates of $C_1(\nu)$ and $C_2(\nu)$ respectively. We see that $\phi_\pm$ are not square integrable near the origin when $\nu^2 \geq 1$. In this case, $n_+ = n_- = 0$ and H_r is (essentially) self-adjoint in the domain $D(H_r)$. On the other hand, both $\phi_\pm$ are square integrable when either $-1 < \nu < 0$ or $0 < \nu < 1$. We therefore see that for any value of ν in these ranges, H_r has deficiency indices $(1,1)$. In this case, H_r is not self-adjoint on the domain $D(H_r)$ but admits self-adjoint extensions. The domain $D_z(H_r)$ in which H_r is self-adjoint contains all the elements of $D(H_r)$ together with elements of the form $\phi_+ + e^{iz}\phi_-$, where $z \in R$ (mod 2π). We now proceed to obtain the spectrum of H_r in the domain $D_z(H_r)$.

The solution of the eigenvalue equation for H_r can be written as

$$R(r) = Br^{\frac{1}{2}} H_\nu^{(1)}(qr),$$

where $q^2 = \epsilon$. Note that in the limit $r \to 0$,

$$\phi_+(r) + e^{iz}\phi_-(r) \to \left[C_1(\nu) + e^{iz}C_1^*(\nu)\right]r^{\nu+\frac{1}{2}}$$
$$+ \left[C_2(\nu) + e^{iz}C_2^*(\nu)\right]r^{-\nu+\frac{1}{2}}$$

and

$$R(r) \to D_1(\nu, q)r^{\nu+\frac{1}{2}} + D_2(\nu, q)r^{-\nu+\frac{1}{2}},$$

where $D_1(\nu, q) = \frac{i}{\sin \pi \nu} \frac{e^{-i\pi\nu} q^\nu}{2^\nu \Gamma(1+\nu)}$ and $D_2(\nu, q) = -\frac{i}{\sin \pi \nu} \frac{q^{-\nu}}{2^{-\nu}\Gamma(1-\nu)}$. If $R(r) \in D_z(H_r)$, then the coefficients of $r^{\nu+\frac{1}{2}}$ and $r^{-\nu+\frac{1}{2}}$ in the above equations must match. Comparing these coefficients we get the bound state energy as

$$E = -\frac{\hbar^2}{2\mu} \left[\cos \frac{\pi\nu}{2} + \cot(\frac{z}{2} + \frac{\pi\nu}{4}) \sin \frac{\pi\nu}{2} \right]^{\frac{1}{\nu}} .$$

Thus we see that for a given value of ν within the allowed range, H_r admits a single bound state with energy E, provided the quantity within the brackets in the expression for E is positive. Keeping in mind that ϵ is negative, we see that the the bound state eigenfunction is given by

$$R(r) = Br^{\frac{1}{2}}H_\nu^{(1)}(i\sqrt{|\epsilon|}r),$$

where B is the normalization constant. The bound state energy and the eigenfunction depends on the choice of the self-adjoint extension parameter z, which classifies the inequivalent boundary conditions.

The case for $\nu = 0$ or $\lambda = -\frac{1}{4}$ can be handled in a similar fashion. The bound state energy and the wave function in this case are given by

$$E = -\frac{\hbar^2}{2\mu}\exp \left[\frac{\pi}{2}\cot\frac{z}{2} \right] ,$$

and

$$\psi(r) = \sqrt{-2\epsilon r}K_0 \left(\sqrt{-\epsilon r} \right)$$

respectively, where K_0 is the modified Bessel function.

We now come to the important issue of critical dipole moment of polar molecules, which is required to bind an electron. The above analysis shows that for $0 \le \nu < 1$, the radial Hamiltonian describing an electron in the field of a polar molecule admits a single bound state. This implies that if constant λ is in the range $-1/4 \le \lambda < 3/4$, the system admits a bound state. Now recall that λ must be negative. Combining these, we can conclude that for any real value of λ such that $-\frac{1}{4} \le \lambda < 0$, the system describing an electron in a dipole field admits a single bound state. Thus the mathematical analysis suggests that any molecule with a non-zero but arbitrarily small dipole moment may be able to capture an electron to form a bound state. In particular, this argument shows that molecules such as H_2S and HCl, whose dipole moments are smaller than the critical dipole moment obtained from the usual analysis can also capture electrons. This is consistent with the experimentally observed anomalous electron scattering in these molecules. The exact numerical value of the bound state energy would depend on the choice of the self-adjoint extension parameter z which characterizes the boundary conditions at the origin. This gives a one-parameter family of inequivalent quantizations The figure shows the

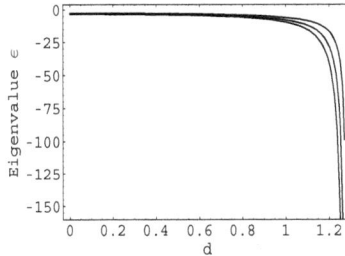

Fig. 4.1 A plot of binding energy of electron as a function of $d = \frac{2\mu e L}{\hbar^2}$ of the polar molecules for three different values of the self-adjoint extension parameter z. From top to bottom $z = \frac{\pi}{10}, \frac{\pi}{8}, \frac{\pi}{6}$ respectively.

plot of the binding energy as a function of the dipole moment d. It is clear that molecules with arbitrary small dipole moment can bind electron, and that the bound state energies may also be very small.

Let us now discuss the physical meaning of the self-adjoint extension parameter z in the context of this model. In our analysis we only consider the potential of the electron due to the dipole effect of the polar molecules. But in reality a polar molecule could have higher moments or could also have short range interactions. These interactions cannot be directly incorporated in the Schrodinger's equation. The method of self-adjoint extension captures the cumulative effect of all such interactions and encodes them in a single real parameter z. The exact value of z cannot be determined from our theory and has to be obtained empirically.

We now address the issue of breaking of scale invariance due to quantization in this system, leading to a quantum mechanical anomaly. This feature is crucial for the formation of any bound state in this system, which is classically scale invariant. Consider the case for $\nu \neq 0$. The anomaly arises as the scaling operator $\Lambda = \frac{-i}{2}(r\frac{d}{dr} + \frac{d}{dr}r)$ acting on an arbitrary element $\phi \in D_z(H_r)$, takes the wavefunction out of the domain of the

Hamiltonian. This can be seen as follows. In the limit $r \to 0$, we have

$$\Lambda\phi(r) \to \frac{(1+\nu)}{i} \left[C_1(\nu) + e^{iz} C_1^*(\nu) \right] r^{\nu + \frac{1}{2}}$$

$$+ \frac{(1-\nu)}{i} \left[C_2(\nu) + e^{iz} C_2^*(\nu) \right] r^{-\nu + \frac{1}{2}}$$

In order for $\Lambda\phi(r) \in D_z(H_r)$, we must have $\Lambda\phi(r) \sim C\phi(r)$ where C is a constant. However, the two terms on the right hand side of the above equation are multiplied by two different factors, i.e. $(1 + \nu)$ and $(1 - \nu)$. Due to the presence of these different multiplying factors, we see that $\Lambda\phi(r)$ in general does not belong to $D_z(H_r)$. Scale invariance is thus broken at the quantum level for generic values of z, due to the choice of the domains of self-adjointness. Since the domains encode the boundary conditions, which in turn capture the effects of the short-range interactions in the polar molecules, we can qualitatively say that the short distance physics is responsible for breaking the scale invariance. From the above equation it is clear that for special choice of $z = -\frac{\nu\pi}{2}$ and $z = -\frac{3\nu\pi}{2}$, $\Lambda\phi(r) \in D_z(H_r)$ and the scaling symmetry is recovered. For these choices of z, the bound states do not exist. A similar analysis can be performed when $\nu = 0$ as well.

4.2.5   *Calogero Model with Confining Interaction*

As a final example of the method of von Neumann, we shall discuss some unusual solutions of a Calogero models, which is a type of integrable models. The rational Calogero model, which we analyze below, is described by N identical particles interacting with each other through a long-range inverse-square and harmonic interaction on the line. This is one of the most well studied, exactly solvable many-particle quantum mechanical systems. This model and its variants have applications in many branches of theoretical physics.

The Hamiltonian of the rational Calogero model is given by

$$H = -\sum_{i=1}^{N} \frac{\partial^2}{\partial x_i^2} + \sum_{i \neq j} \left[\frac{a^2 - \frac{1}{4}}{(x_i - x_j)^2} + \frac{\Omega^2}{16} (x_i - x_j)^2 \right],$$

where a, Ω are constants, x_i is the coordinate of the ith particle and units have been chosen such that $2m\hbar^{-2} = 1$. We are interested in finding normalizable solutions of the eigenvalue problem

$$H\psi = E\psi.$$

To that end, consider the above eigenvalue equation in a sector of configuration space corresponding to a definite ordering of particles given by $x_1 \geq x_2 \geq \cdots \geq x_N$. The eigenfunctions of the Hamiltonian H that are translation-invariant can be written as

$$\psi = \prod_{i<j} (x_i - x_j)^{a+\frac{1}{2}} \; \phi(r) \; P_k(x),$$

where $x \equiv (x_1, x_2, \ldots, x_N)$,

$$r^2 = \frac{1}{N} \sum_{i<j} (x_i - x_j)^2$$

and $P_k(x)$ is a translation-invariant as well as homogeneous polynomial of degree $k(\geq 0)$ which satisfies the equation

$$\left[\sum_{i=1}^{N} \frac{\partial^2}{\partial x_i^2} + \sum_{i \neq j} \frac{2(a + \frac{1}{2})}{(x_i - x_j)} \frac{\partial}{\partial x_i} \right] P_k(x) = 0.$$

The existence of complete solutions of this equation was discovered by Calogero. Using the above equations we get

$$H_r \phi = E \phi,$$

where

$$H_r = \left[-\frac{d^2}{dr^2} - (1 + 2\nu) \frac{1}{r} \frac{d}{dr} + w^2 r^2 \right]$$

with $w^2 = \frac{1}{8} \Omega^2 N$ and

$$\nu = k + \frac{1}{2}(N - 3) + \frac{1}{2} N(N - 1)(a + \frac{1}{2}).$$

H_r is the effective Hamiltonian in the "radial" direction. Note that that $\phi(r) \in L^2[R^+, d\mu]$, where the measure is given by $d\mu = r^{1+2\nu} dr$.

The Hamiltonian H_r is a symmetric operator on the domain $D(H_r) \equiv \{\phi(0) = \phi'(0) = 0, \; \phi, \; \phi' \text{ absolutely continuous}\}$. To determine whether H_r is self-adjoint in $D(H_r)$, we have to first look for square integrable solutions of the equations

$$H_r^* \phi_\pm = \pm i \phi_\pm,$$

where H_r^* is the adjoint of H_r. Note that H_r^* is given by the same differential operator as H_r although their domains might be different. The solutions are given by

$$\phi_\pm(r) = e^{-\frac{wr^2}{2}} U\left(d_\pm, c, wr^2\right),$$

where $d_\pm = \frac{1+\nu}{2} \mp \frac{i}{4w}$, $c = 1+\nu$ and U denotes the confluent hypergeometric function of the second kind. The asymptotic behaviour of U together with the exponential factor in the solutions ensures that $\phi_\pm(r)$ vanish at infinity. These solution have different short distance behaviour for $\nu \neq 0$ and $\nu = 0$. From now onwards, we shall restrict our discussion to the case for $\nu \neq 0$. The analysis for $\nu = 0$ has been left as an exercise. When $\nu \neq 0$, $U(d_\pm, c, wr^2)$ can be written as

$$U\left(d_\pm, c, wr^2\right) = C \left[\frac{M\left(d_\pm, c, wr^2\right)}{\Gamma(b_\pm)\Gamma(c)} - \left(wr^2\right)^{1-c} \frac{M\left(b_\pm, 2-c, wr^2\right)}{\Gamma(d_\pm)\Gamma(2-c)} \right],$$

where $b_\pm = \frac{1-\nu}{2} \mp \frac{i}{4w}$, $C = \frac{\pi}{sin(\pi+\nu\pi)}$ and M denotes the confluent hypergeometric function of the first kind. In the limit $r \to 0$, $M(d_\pm, c, wr^2) \to 1$. Thus as $r \to 0$, we get

$$|\phi_\pm(r)|^2 d\mu \to \left[A_1 r^{(1+2\nu)} + A_2 r + A_3 r^{(1-2\nu)} \right] dr,$$

where A_1, A_2 and A_3 are constants independent of r. It is now clear that in the limit $r \to 0$, the functions $\phi_\pm(r)$ are not square-integrable if $| \nu | \geq 1$. In that case, $n_+ = n_- = 0$ and H_r is essentially self-adjoint in the domain $D(H_r)$. However, if either $0 < \nu < 1$ or $-1 < \nu < 0$, the functions $\phi_\pm(r)$ are indeed square-integrable. Thus if ν lies in these ranges, we have $n_+ = n_- = 1$ and Hamiltonian H_r is not self-adjoint in $D(H_r)$ but admits self-adjoint extensions. The domain $D_z(H_r)$ in which H_r is self-adjoint contains all the elements of $D(H)$ together with elements of the form $\phi_+ + e^{iz}\phi_-$, where $z \in R$ (mod 2π). We can similarly show that $n_+ = n_- = 1$ for $\nu = 0$ as well. Thus the self-adjoint extensions of this model exist when $-1 < \nu < 1$.

The range of ν required for the existence of the self-adjoint extension implies that for given values of N and k, $a + \frac{1}{2}$ must lie on the range

$$-\frac{N-1+2k}{N(N-1)} < a + \frac{1}{2} < -\frac{N-5+2k}{N(N-1)}.$$

For $N \geq 3$, we have various types of boundary conditions depending on the value of the parameter $a + \frac{1}{2}$. Consider first the case with $a + \frac{1}{2} \geq \frac{1}{2}$. This corresponds to the boundary condition considered by Calogero for which both the wave-function and the current vanish as $x_i \to x_j$. In this case, $\nu > 1$ for all values of $k \geq 0$. The corresponding Hamiltonian is essentially self-adjoint in the domain $D(H_r)$, leading to a unique quantum theory. The next case is described by the condition $0 < a + \frac{1}{2} < \frac{1}{2}$. Here, The wave-function vanishes in the limit $x_i \to x_j$, though the current may show a divergent behaviour in the same limit. In this case ν is positive and k

must be equal to zero so that ν may belong to the range $0 < \nu < 1$. The corresponding constraint on $a + \frac{1}{2}$ is given by $0 < a + \frac{1}{2} < \frac{5-N}{N(N-1)}$, which can only be satisfied for $N = 3$ and 4. Finally, we have the condition $-\frac{1}{2} < a + \frac{1}{2} < 0$. The lower bound on $a + \frac{1}{2}$ is obtained from the condition that the wavefunction be square-integrable. The parameter $a + \frac{1}{2}$ in this range leads to a singularity in the wavefunction resulting from the coincidence of any two or more particles. Using permutation symmetry, such an eigenfunction can be extended to the whole of configuration space, although not in a smooth fashion. The new quantum states in this case exist for arbitrary N and even for non-zero values of k. In fact, imposing the condition that the upper bound on $a + \frac{1}{2}$ should be greater than $-\frac{1}{2}$, we get that k is restricted as $k < \frac{1}{4}\left(N^2 - 3N + 10\right)$. It can also be shown that there are only two allowed values of k when both N and $a + \frac{1}{2}$ are kept fixed.

In order to determine the spectrum we note that the solution of the radial equation which is bounded at infinity is given by

$$\phi(r) = Be^{-\frac{wr^2}{2}}U(d, c, wr^2),$$

where $d = \frac{1+\nu}{2} - \frac{E}{4w}$ and B is a constant. In the limit $r \to 0$,

$$\phi(r) \to BC\left[\frac{1}{\Gamma(b)\Gamma(c)} - \frac{w^{-\nu}r^{-2\nu}}{\Gamma(d)\Gamma(2-c)}\right],$$

where $b = \frac{1-\nu}{2} - \frac{E}{4w}$. On the other hand, as $r \to 0$,

$$\phi_+ + e^{iz}\phi_- \to C\left[\frac{1}{\Gamma(c)}\left(\frac{1}{\Gamma(b_+)} + \frac{e^{iz}}{\Gamma(b_-)}\right) - \frac{w^{-\nu}r^{-2\nu}}{\Gamma(2-c)}\left(\frac{1}{\Gamma(d_+)} + \frac{e^{iz}}{\Gamma(d_-)}\right)\right].$$

If $\phi(r) \in D_z(H)$, then the coefficients of different powers of r in the above limits must match. Comparing the coefficients of the constant term and $r^{-2\nu}$ we get

$$f(E) \equiv \frac{\Gamma\left(\frac{1-\nu}{2} - \frac{E}{4w}\right)}{\Gamma\left(\frac{1+\nu}{2} - \frac{E}{4w}\right)} = \frac{\xi_2\cos(\frac{z}{2} - \eta_1)}{\xi_1\cos(\frac{z}{2} - \eta_2)},$$

where $\Gamma\left(\frac{1+\nu}{2} + \frac{i}{4w}\right) \equiv \xi_1 e^{i\eta_1}$ and $\Gamma\left(\frac{1-\nu}{2} + \frac{i}{4w}\right) \equiv \xi_2 e^{i\eta_2}$. This is our energy equation. For given values of the parameters ν and w, the bound state energy E is obtained as a function of z. Different choices of z thus leads to inequivalent quantizations of the many-body Calogero model. Moreover we see that for fixed value of z, the Calogero model with parameters (w, ν) and $(w, -\nu)$ produces identical energy spectrum although the corresponding wavefunctions are different.

We can get an analytical value of the spectrum when the right hand side of the energy equation is either 0 or ∞. When the right hand side is 0, we

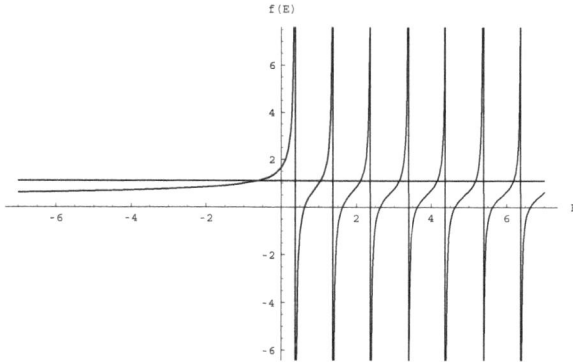

Fig. 4.2   A plot of the energy equation using Mathematica with $w = 0.25$, $\nu = 0.25$ and $z = -1.5$. The horizontal straight line corresponds to the value of the right hand side of the energy equation.

must have the situation where $\Gamma\left(\frac{1+\nu}{2} - \frac{E}{4w}\right)$ blows up, or, $E_n = 2w(2n + \nu + 1)$ where n is a positive integer. This happens for the special choice of $z = z_1 = \pi + 2\eta_1$. These eigenvalues and the corresponding eigenfunctions are analogous to those found by Calogero although for a different parameter range. Similarly, when the right hand side of the energy equation is ∞, an analysis similar to the one above shows that $E_n = 2w(2n - \nu + 1)$. This happens for the special value of z given by $z = z_2 = \pi + 2\eta_2$.

For choices of z other than z_1 or z_2, the nature of the spectrum can be understood from Fig. 4.2, which is a plot of the energy equation for specific values of ν, z and w. In that plot, the horizontal straight line corresponds to the right hand side of that equation. The energy eigenvalues are obtained from the intersection of $f(E)$ with the horizontal straight line. Note that the spectrum generically consists of infinite number of positive energy solutions and at most one negative energy solution. The existence of the negative energy states can be understood in the following way. For large negative values of E, the asymptotic value of $f(E)$ is given by $\left(\frac{E}{4w}\right)^{-\nu}$, which monotonically tends to 0 or $+\infty$ for $\nu > 0$ or $\nu < 0$ respectively. When $\nu > 0$, the negative energy state will exist provided right hand side of the energy equation lies between 0 and $\frac{\Gamma(\frac{1-\nu}{2})}{\Gamma(\frac{1+\nu}{2})}$. Similarly, when $\nu < 0$, the negative energy state will exist when the right hand side of the energy equation lies between $\frac{\Gamma(\frac{1-\nu}{2})}{\Gamma(\frac{1+\nu}{2})}$ and $+\infty$. For any given values of ν and w, the position of the horizontal straight line in Fig. 4.2 can always be adjusted to lie anywhere between $-\infty$ and $+\infty$ by suitable choices of z. Thus the spectrum would always contain a negative energy state for some choice of the parameter z.

An important aspect of the spectrum obtained here is that the spectrum is not equispaced for finite values of E and for generic values of z. For example, it is seen from the energy equation that the ratio

$$\frac{f(E + 4w)}{f(E)} = \frac{\frac{E}{4w} + \frac{1-\nu}{2}}{\frac{E}{4w} + \frac{1+\nu}{2}}$$

in general is not unity except when $E \to \infty$. This may seem surprising with the presence of $SU(1,1)$ as the spectrum generating algebra in this system, which demands that the eigenvalues be evenly spaced. In order to address this issue, we consider the action of the dilatation generator $D = \frac{1}{2}\left(r\frac{d}{dr} + \frac{d}{dr}r\right)$ on an element $\phi(r) = \phi_+(r) + e^{iz}\phi_-(r)$. In the limit $r \to 0$, we have

$$D\phi = \frac{C}{2}\left[\frac{1}{\Gamma(c)}\left(\frac{1}{\Gamma(b_+)} + \frac{e^{iz}}{\Gamma(b_-)}\right) - \frac{r^{-2\nu}(1-4\nu)}{\Gamma(2-c)}\left(\frac{1}{\Gamma(d_+)} + \frac{e^{iz}}{\Gamma(d_-)}\right)\right].$$

We therefore see that $D\phi(r) \in D_z(\tilde{H})$ only for $z = z_1$ or $z = z_2$. Thus the generator of dilatations does not in general leave the domain of the Hamiltonian invariant. Consequently, $SU(1,1)$ cannot be implemented as the spectrum generating algebra except for $z = z_1, z_2$.

For $N \geq 3$, the range of $a + \frac{1}{2}$ for which the new quantum states have been found is different from what was used by Calogero. The $N = 2$ Calogero model however admits new quantum states even in the range of $a + \frac{1}{2}$ considered by Calogero. When $N = 2$, k must be equal to zero and we get $\nu = a$. In this case, the system therefore admits self-adjoint extensions and new quantum states when $-1 < a < 1$. The eigenvalue problem for $N = 2$ was solved by Calogero with the condition that $a > 0$. Thus when $0 < a < 1$, this analysis predicts a larger family of solutions labelled by the parameter z. This set of solutions reduces to that found by Calogero for $z = z_1$.

Further Reading and Selected References

M. Reed and B. Simon, Methods of Modern Mathematical Physics, volume 1 and 2, (Academic Press, New York). This is a standard textbook on self-adjoint extension.

V. Hutson and J.S. Pym, Applications of Functional Analysis and Operator Theory, (Academic Press, London, 1980). Contains a good exposition of operator theory and self-adjoint extensions.

S. Albeverio et al, Solvable Models in Quantum Mechanics, AMS Chelsea Publishing, 2004. This book has several solved examples of self-adjoint extension.

R. Jackiw in M.A.B. Beg Memorial Volume, A. Ali and P. Hoodbhoy, eds. (World Scientific, Singapore, 1991). Discusses self-adjoint extension of the delta function potential and relates that to renormalization group theory.

E. Farhi and S. Gutmann, Int. J. Mod. Phys. **A5**, 3029, 1990 contains a discussion of self-adjoint extension in the context of path integrals.

P. Gerbert, Phys. Rev. **D 40**, 1346 (1989) discusses self-adjoint extension of fermions in the presence of cosmic strings.

J. G. Esteve, Phys. Rev. **D 34**, 674 (1986); J. G. Esteve, Phys. Rev. **D 66**, 125013 (2002); A.P. Balachandran and Amilcar R. de Queiroz Phys. Rev. **D 85** (2012) 025017; Kumar S. Gupta and Amilcar de Queiroz, Mod. Phys. Lett. **A 29** (2014) 13, 1450064. These papers discuss self-adjoint extension and its relation to symmetry breaking and anomaly.

A good discussion on electron scattering by polar molecules is given in the paper by J-M. Levy Leblond, Phys. Rev. **153**, 1 (1967).

For more details on the effect of self-adjoint extension on the anomalous electron scattering by polar molecules, see the paper by Pulak Ranjan Giri, Kumar S. Gupta, S. Meljanac and A. Samsarov, Phys. Lett. A372 (2008) 2967-2970 and references therein. Figure 4.1 is reprinted from this paper with permission from Elsevier.

For more details of Calogero models and self-adjoint extension, see the following papers : B. Basu-Mallick et al Phys. Lett. A311 (2003) 87-92; Nucl. Phys. B659 (2003) 437-457; Eur. Phys. J. C58 (2008) 159-170; S. Meljanac et al, Eur. Phys. J. C49 (2007) 875-889.

For a discussion of boundary conditions and self-adjoint extension related to some condensed matter systems see M. Asorey et al, JHEP 1312 (2013) 073.

Chapter 5

Electronic Properties of Graphene

5.1 Introduction

Graphene is one of the wonder materials of nature. Planar graphene consists of a two dimensional layer of carbon atoms arranged in a honeycomb lattice, which was experimentally fabricated in the laboratory in 2004. One of the remarkable properties of graphene is that the low energy excitations in a graphene monolayer obey a 2D massless Dirac equation. The theoretical derivation of this result is based on the nearest neighbour hopping and ignores the electron-electron interactions in graphene. The relativistic quantum Hall effect in graphene has clearly demonstrated the existence of Dirac like quasiparticles. The quasiparticles in pristine graphene are massless. Under suitable conditions the Dirac excitations could be massive, in which case the graphene system is gapped. In either case, the quasiparticles in graphene are negatively charged and carry the same electric charge as that of an electron.

The main focus of our discussion would be the quantum dynamics of the Dirac type quasiparticles in graphene in the presence of a external Coulomb type charge impurity. At the level of the Dirac equation, this system is then governed by two parameters, namely the mass m of the quasiparticles and the coupling β of the external charge impurity with the quasiparticles. The other important ingredient is provided by the boundary condition that the Dirac equation satisfies. The solutions of the Dirac equation will depend on m, β and on the imposed boundary condition. Before going into mathematical details, we first discuss certain qualitative aspects of this system.

The most natural choice for graphene is $m = 0$, when the system is gapless. In 2D, massless Dirac excitations do not form any bound state.

This is generally attributed to the Klein tunneling. For such a system we have only scattering states. However, when $m \neq 0$ and the system is gapped, in the presence of suitable interactions, the quasiparticles spectrum admits both bound and scattering states.

Let us now discuss the qualitative implications of the coupling β, which plays the role of the fine structure constant in graphene. Consider an external charge impurity of strength Ze, where e is the electron charge and Z is a constant. This impurity will interact with the negatively charged quasiparticles of graphene, which also carry a charge e. The effective fine structure constant has the form $\beta = \frac{Ze^2}{\kappa v_F}$, where v_F is the Fermi velocity in graphene and κ is the effective dielectric constant which is approximately equal to 2.9. In the usual electromagnetic interaction, v_F is replaced with c, which is the velocity of light. What makes graphene very interesting is the fact that the Fermi velocity V_F is graphene is $\sim 10^6$ meters/second, which is 300 times less than the velocity of light. Thus, for any given external charge Ze, the effective coupling β is graphene is approximately 300 times stronger than electromagnetic fine structure constant $\alpha \sim \frac{1}{137}$. Thus for relatively low values of the external charge $Ze \sim 1$, the coupling $\beta \sim 1$, which is the strong coupling regime. Contrast this with the situation in nuclear physics, where one needs super-heavy nuclei with $Z \geq 170$ to achieve at the strong coupling. Such nuclei are unstable and hence it is difficult to do experiments with those. On the contrary, graphene provides a relatively easy way to reach the strong coupling regime. Why is the strong coupling regime interesting? In this regime the nonperturbative effects of QED are supposed to become important. It is expected that the quantum vacuum would break down and what Landau famously called as *fall to the centre effect* would be observed. In particular, it has been predicted that in the strong QED limit, there would be a number of quasi-bound states for graphene. Such states have been predicted to exist both numerically and analytically and now there are strong experimental indications of their existence in graphene systems. Graphene therefore provides a unique opportunity to study strong nonperturbative QED effects in a simple laboratory setup.

Finally let us talk about the boundary conditions. Note that the Dirac equation in graphene is obtained at low energies or equivalently in the long wavelength limit. Now, once a charge impurity is introduced in graphene, it can have various local short distance interactions with the other charged particles in the system. Within the framework of the Dirac description, which is valid only at low energies, there is no way to introduce such interactions directly in the Hamiltonian. In quantum theory, it is known that interactions can be either included in the Hamiltonian or they may be coded in the boundary conditions. For example, in the Bohm-Aharonov effect, one can either have a free Hamiltonian with a twisted boundary condition on the electron wavefunctions or can have a Hamiltonian with the vector potential but with a simple boundary condition. In every system such a one-to-one correspondence between may not exist. For graphene with a charge impurity, the collective effect of the short distance interactions can however be encoded in the boundary conditions. Such boundary conditions are obtained by demanding the conservation of probability current, or a unitary time evolution. The method of self-adjoint extension due to von Neumann, which has been discussed before, would be used to obtain all such allowed boundary conditions.

5.2 Tight-Binding Model and the Dirac Equation

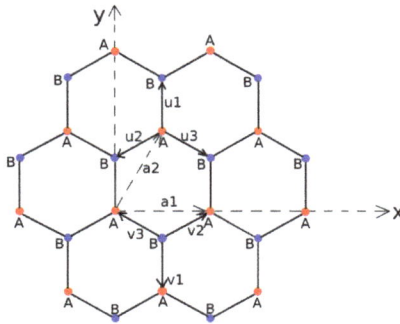

Fig. 5.1 Lattice structure of graphene.

Graphene is made out of carbon atoms arranged in a hexagonal structure as shown in Fig. 5.1. The bravais lattice is triangular and the unit cell contains two types of sites, which are usually denoted by A and B. The points belonging to the A sites are obtained by linear combinations of the basis vectors $\vec{a_1}$ and $\vec{a_2}$,

$$\vec{A}(n_1, n_2) = n_1\vec{a_1} + n_2\vec{a_2}.$$

where n_1 and n_2 are positive integers. In terms of the unit vectors $\hat{e}_x$ and $\hat{e}_y$, we have

$$\vec{a_1} = \sqrt{3}a\ \hat{e}_x, \qquad \vec{a_2} = \frac{\sqrt{3}}{2}a\ \hat{e}_x + \frac{3}{2}a\ \hat{e}_y.$$

The two sublattices are connected by

$$\vec{u_1} = a\,\hat{e}_y, \quad \vec{u_2} = -\frac{\sqrt{3}}{2}a\,\hat{e}_x - \frac{a}{2}\,\hat{e}_y, \quad \vec{u_3} = \frac{\sqrt{3}}{2}a\,\hat{e}_x - \frac{a}{2}\,\hat{e}_y$$

Similarly, the points in the B sites are given by

$$\vec{B}(m_1, m_2) = m_1\,\vec{a_1} + m_2\,\vec{a_2} + \vec{u_1}.$$

In order to define the reciprocal lattice, let us first denote $\vec{a_3} = \hat{e}_z$. Then, the reciprocal basis vectors can be written as Then the reciprocal lattice basis vectors are given by

$$\vec{K_1} = 2\pi \frac{\vec{a_2} \times \vec{a_3}}{\vec{a_1} \cdot (\vec{a_2} \times \vec{a_3})} = \frac{2\pi}{\sqrt{3}a}\,\hat{e}_x - \frac{2\pi}{3a}\,\hat{e}_y$$

and

$$\vec{K_2} = 2\pi \frac{\vec{a_3} \times \vec{a_1}}{\vec{a_1} \cdot (\vec{a_2} \times \vec{a_3})} = \frac{4\pi}{3a}\,\hat{e}_y.$$

The graphene lattice is monatomic with the sites occupied by the carbon atoms. If the sublattice symmetry is broken due to some reason, the energies of electrons localized at sites A and B would be different. This difference is parametrized by ζ. We consider only the nearest neighbour interaction. The Hamiltonian is then given by

$$H = \gamma \sum_{\vec{A},i}[U^\dagger(\vec{A})V(\vec{A} + \vec{u_i}) + V^\dagger(\vec{A} + \vec{u_i})U(\vec{A})]$$

$$+\zeta \sum_{\vec{A}}[U^\dagger(\vec{A})U(\vec{A}) - V^\dagger(\vec{A} + \vec{u_1})V(\vec{A} + \vec{u_1})].$$

where $U^\dagger$ and U ($V^\dagger$ and V) are the creation and destruction operators for electrons localized on sites A and B respectively and γ is the hopping parameter. As mentioned before, the difference in energies of the electrons localized at A and B is denoted by ζ. In what follows, we ignore the electron spin. Using the Fourier transform

$$U(\vec{A}) = \int_B \frac{d^2k}{(2\pi)^2}\,e^{i\vec{k}\cdot\vec{A}}U(\vec{k}) \quad \text{and} \quad V(\vec{B}) = \int_B \frac{d^2k}{(2\pi)^2}\,e^{i\vec{k}\cdot\vec{B}}V(\vec{k}),$$

the Hamiltonian takes the form

$$H = \int_B \frac{d^2k}{(2\pi)^2}\left(U^\dagger(\vec{k})\ V^\dagger(\vec{k})\right)\begin{pmatrix} \zeta & \gamma\sum_i e^{i\vec{k}\cdot\vec{u_i}} \\ \gamma\sum_i e^{-i\vec{k}\cdot\vec{u_i}} & -\zeta \end{pmatrix}\begin{pmatrix} U(\vec{k}) \\ V(\vec{k}) \end{pmatrix}.$$

The energy eigenvalue λ can be determined from the equation

$$\begin{vmatrix} \zeta - \lambda & \gamma\sum_i e^{i\vec{k}\cdot\vec{u_i}} \\ \gamma\sum_i e^{-i\vec{k}\cdot\vec{u_i}} & -\zeta - \lambda \end{vmatrix} = 0,$$

which gives

$$\lambda = \pm \sqrt{\zeta^2 + \gamma^2 |\sum_i e^{i\vec{k}\cdot\vec{u}_i}|^2} \ .$$

The positive and negative values of λ correspond to the conduction and valence bands respectively. The separation between the bands is minimal when

$$|\sum_i e^{i\vec{k}\cdot\vec{u}_i}| = 0.$$

This equation is satisfied for six points in the momentum space given by

$$k_x = \pm \frac{4\pi}{3\sqrt{3}a}, \qquad k_y = 0 \ ;$$

$$k_x = \pm \frac{2\pi}{3\sqrt{3}a}, \qquad k_y = \pm \frac{2\pi}{3a}.$$

These points define the six vertices of the hexagonal Brillouin zone, where the dispersion is linear. They are known as the Dirac points. Out of the six, there are only two inequivalent Dirac points, the rest being related by lattice symmetry. What is so special about these Dirac points? To see that, consider a small fluctuation $\delta\vec{k}$ around one them, which has $k_x = \frac{2\pi}{3\sqrt{3}a}$, $k_y = 0$. We write

$$\vec{k} = \frac{4\pi}{3\sqrt{3}a}\hat{e}_x + \delta\vec{k}.$$

Let us now consider the structure of the Hamiltonian around this Dirac point. Assuming that the momentum fluctuation is small and in the continuum limit, the Hamiltonain around this Dirac point looks like

$$= -v_F[\vec{\sigma}\cdot\delta\vec{k} - mv_F\sigma_3],$$

where $mv_F = \frac{2\zeta}{3\gamma}$ and $v_F = \frac{3\gamma a}{2}$ is the Fermi velocity. In planar graphene $v_F \approx 1 \times 10^6 m/s$. The Pauli matrices $\sigma_i, i = 1, 2, 3$ act on the two sublattice indices which define the two components of the Dirac wavefunction of graphene. In position space, the above Hamiltonian takes the form

$$H_0 = -i\hbar v_F(\tau_3\sigma_1\partial_x + \sigma_2\partial_y) + \Delta\sigma_3 \ .$$

Here the term $\tau_3 = \pm 1$ distinguishes between the Dirac points and $\Delta \equiv mv_F{}^2$. At low energies, we neglect the scattering between the Dirac points and hence we can set $\tau_3 = +1$ without loss of generality. By suitable choice of units it is possible to set $\hbar = v_F = 1$. This choice shall be called natural units. We shall keep these constants explicitly when they are necessary to illustrate some important physics aspetcs. In the natural units, the above Hamiltonian can be obtained from the Lagrangian

$$L_0 = \bar{\psi}(i\gamma^\mu \partial_\mu - m)\psi, \quad \bar{\psi} \equiv \psi^\dagger \gamma^0,$$

where $\bar{\psi} \equiv \psi^\dagger \gamma^0$. and $\gamma_0 = \sigma_3, \gamma^1 = i\sigma_2$ and $\gamma_2 = -i\sigma_1$.

What have we learnt so far? We started with a nonrelativistic system of electrons in planar graphene with a hopping interaction γ between nearest neighbours and a parameter ζ that characterized the difference of electron energies among the sites of the two sublattices of graphene. How could such a system be governed by a Dirac equation? The key assumptions were nearest neighbour interactions, small momenta or long wavelength of the quasiparticles around the Dirac points and the continuum limit. With these assumptions, the low energy quasiparticles around the Dirac points in graphene obey a Dirac equation. Recall that we have ignored the electron spin. Thus the two components of the Dirac wavefunction corresponds to the sublattice indices and not to the electron spins.

For the rest of this section, we shall summarize the properties of this free planar Dirac equation in a polar coordinate system. This coordinate system is adapted to the symmetries of the problem in the presence of a Coulomb charge or a topological defect, which we shall study later. Let us start with the eigenvalue problem

$$H_0 \Psi = E_0 \Psi.$$

Let $\hat{l}_z = -i(x\partial_y - y\partial_x) = -i\partial_\phi$ denote the component of the angular momentum perpendicular to the graphene sheet, which is assumed to lie on the $x-y$ plane. We denote the eigenvalues of $\hat{l}_z$ by l_z, which is an integer. Note that $[H_0, \hat{l}_z] \neq 0$, which means that the eigenstates cannot be labelled by the eigenvalues of $\hat{l}_z$. Instead we find that $[H_0, \hat{l}_z + \frac{1}{2}\hat{\sigma}_z] = 0$. Thus the eigenvalues of the operator $\hat{l}_z + \frac{1}{2}\hat{\sigma}_z$ can be used to label the eigenstates of H_0. The eigenvalue of $\hat{\sigma}_z$ is chosen as 1. We define $j = l_z + \frac{1}{2}$. In the radial polar coordinates (r, θ), the eigenstates have the form

$$\Psi(r, \phi) = \begin{pmatrix} f(r)\, \Phi_{j-\frac{1}{2}}(\phi) \\ ig(r)\, \Phi_{j+\frac{1}{2}}(\phi) \end{pmatrix},$$

where $\Phi_j(\phi) = e^{ij\phi}$, $E_0 = \pm\sqrt{p^2 + m^2}$ and $p = \sqrt{p_x^2 + p_y^2}$. Note that p_x and p_y are the eigenvalues of the momentum operators $-i\partial_x$ and $-i\partial_y$ respectively. For $|E_0| > m$. the radial equations are given by

$$-\frac{1}{r}\frac{d}{dr}\left(r\frac{df}{dr}\right) + \frac{(j-\frac{1}{2})^2}{r^2}f = p^2 f,$$

$$-\frac{1}{r}\frac{d}{dr}\left(r\frac{dg}{dr}\right) + \frac{(j+\frac{1}{2})^2}{r^2}g = p^2 g.$$

These are Bessel equations whose solutions are given by the Bessel function $J(pr)$ which is regular at $r = 0$. Using the normalization condition $\int d^2\mathbf{r}\psi_{pj}^\dagger \psi_{p'j'} = 2\pi\delta_{jj'}(p - p')$, we finally get

$$\Psi_{pj}(r,\phi) = \frac{\sqrt{2\pi p}}{\sqrt{2|E_0|}}\left(\begin{array}{c}\sqrt{|E_0 + m|}J_{j-\frac{1}{2}}(pr)\ \Phi_{j-\frac{1}{2}}(\phi)\\ \pm i\sqrt{|E_0 - m|}J_{j+\frac{1}{2}}(pr)\ \Phi_{j+\frac{1}{2}}(\phi)\end{array}\right).$$

In the above discussion, the mass m of the quasiparticles in graphene has been kept arbitrary, except that $m \geq 0$. When $m = 0$ the system is gapless, which is experimentally the most prevalent form of graphene. When $m \neq 0$ but positive, the graphene is said to be gapped. In order for the gapped graphene to form, the degeneracy between the sublattices A and B has to be broken due to some physical effects. In the presence of interactions or defects, the mass m affects the dynamics in a significant fashion. We shall see this in the course of our discussion.

5.3 Gapless Graphene with Coulomb Charge Impurities

We have seen that the dynamics of the low energy quasiparticles in graphene is governed by a 2D Dirac equation. These quasiparticles are negatively charged and carry the same charge as that of an electron. Suppose we introduce a positively charged external Coulomb impurity in the system. Let the strength of this charge impurity be Ze, where Z is a constant and e is the unit electric charge, whose value is given by the magnitude of the charge of an electron. The external charge impurity will interact with the negatively charged quasiparticles in graphene and will contribute an interaction term in the Dirac Hamiltonian which is given by $V(r) = -\frac{Ze^2}{\kappa r}$, where κ being the effective dielectric constant in graphene. The Hamiltonian of the system is given by

$$H = \hbar v_F\left(\begin{array}{cc}0 & -i\partial_x - \partial_y\\ -i\partial_x + \partial_y & 0\end{array}\right) - \frac{Ze^2}{\kappa r},$$

which satisfies the 2D Dirac equation

$$H\Psi = E\Psi,$$

where r is the radial coordinate in the two dimensional $x - y$ plane. The Dirac wavefunction Ψ has two components, which are labelled by the sublattice indices. These indices are not the usual spin indices that appear in the Dirac equation and are sometimes called pseudospin indices. Using the polar coordinates, we can write

$$\Psi(r, \phi) = \begin{pmatrix} F(r) \frac{1}{\sqrt{2\pi}} e^{i(j-\frac{1}{2})\phi} \\ G(r) \frac{1}{\sqrt{2\pi}} e^{i(j+\frac{1}{2})\phi} \end{pmatrix},$$

where ϕ denotes the corresponding polar angle and j is the half integral azimuthal quantum number. Now consider the ansatz

$$F(r) = e^{ikr} r^{\nu-\frac{1}{2}} (u(r) + v(r)),$$
$$G(r) = e^{ikr} r^{\nu-\frac{1}{2}} (u(r) - v(r)),$$

where $\nu = \sqrt{j^2 - \beta^2}$, $\beta \equiv \frac{Ze^2}{\kappa \hbar v_F}$, $k = -\frac{E}{\hbar v_F}$. Note that the quantity β provides a measure of the effective electromagnetic interaction strength in this system. It is analogous to the fine structure constant of QED, with the important difference that the Fermi velocity v_F replaces the velocity of light. Using this ansatz, the radial Dirac operator is obtained as

$$H_r = \begin{pmatrix} r\frac{d}{dr} + \nu + i\beta + 2ikr & -j \\ -j & r\frac{d}{dr} + \nu - i\beta \end{pmatrix}.$$

The Dirac equation now gives rise to two coupled differential equations

$$r\frac{du}{dr} + (\nu + i\beta + 2ikr)u - jv = 0,$$

$$r\frac{dv}{dr} + (\nu - i\beta)v - ju = 0.$$

In terms of the variable $z = -2ikr$, the equation for the variable $v(r)$ can be written as

$$z\frac{d^2v}{dz^2} + (1 + 2\nu - z)\frac{dv}{dz} - (\nu - i\beta)\, v = 0.$$

This shows that v is described by a confluent hypergeometric function and a similar conclusion holds for the function u as well.

Let us pause for a moment and understand the behaviour of the factor $r^{\nu-\frac{1}{2}}$ in the above wavefunction. If $\beta > j$, then ν becomes imaginary. In that case, the quantity $r^{\nu-\frac{1}{2}}$ is a wildly oscillating function near the origin, which is the position of the charge impurity. The minimum value of β for which this happens is called the critical value denoted by β_c. In the channel $j = \frac{1}{2}$, we see that $\beta_c = \frac{1}{2}$. Therefore, if $\beta > \beta_c$, then there is at least one angular momentum channel in which the wavefunction displays this wildly oscillating behaviour near the origin. Such a behaviour indicates a breakdown of the vacuum, which Landau famously called *fall to the centre*. As we shall see below, the quantum dynamics is very different in the sub-critical regime where $\beta < \frac{1}{2}$ as compared to the supercritical regime where $\beta > \beta_c$. We shall first discuss the subcritical regime, where ν is a positive real parameter.

5.3.1 *Boundary Conditions and Self-Adjoint Extensions*

The effective potential due to the charge impurity is given by $-\frac{\beta}{r}$. This is a long range Coulomb interaction that appears in the Dirac equation for graphene. Is this the only effect of the external charge impurity? Recall that graphene is actually a two dimensional array of carbon atoms arranged in a hexagonal lattice. When a charge impurity is introduced in graphene, it can give rise to various short ranged interactions depending on the details of the impurity and the local conditions of the graphene sheet. Such interactions cannot be directly included in the Dirac equation, which is valid only in the long wavelength limit. However, it may be possible to encode the collective effect of all such short range interactions through the choice of suitable boundary conditions. How should we determine the appropriate boundary conditions? Note that in a graphene system with an external charge. the probability current must be conserved. In other words the Hamiltonian must generate a unitary time evolution, or equivalently the Hamiltonian must be self-adjoint. Thus the physical requirement of unitarity leads us to search for all possible boundary conditions that would ensure that the Dirac operator in graphene is self-adjoint. As discussed before, the method of self-adjoint extension due to von Neumann is precisely tailored to achieve this goal and in what follows, we shall apply that method to this problem at hand.

The operator H_r is symmetric (or Hermitian) in the domain $D(H_r)$ defined by

$$D(H_r) = \{\Psi \mid \Psi(0) = \Psi^{'}(0) = 0, \ \Psi, \Psi^{'} \text{ absolutely continuous}, \Psi \in L^2(rdr)\}.$$

The question is if the operator H_r is self-adjoint in $D(H_r)$? Following the method of von Neumann, we consider the equations

$$H_r^{\dagger}\Psi_{\pm} = \pm i\Psi_{\pm},$$

where $H_r^{\dagger}$ is the adjoint of H_r and $\Psi_{\pm}$ are given by

$$\Psi_{\pm} = \begin{pmatrix} (u_{\pm} + v_{\pm}) \frac{1}{\sqrt{2\pi}} e^{i(j-\frac{1}{2})\phi} \\ (u_{\pm} - v_{\pm}) \frac{1}{\sqrt{2\pi}} e^{i(j+\frac{1}{2})\phi} \end{pmatrix} r^{\nu-\frac{1}{2}} e^{\mp r} \equiv \begin{pmatrix} F_{\pm} \frac{1}{\sqrt{2\pi}} e^{i(j-\frac{1}{2})\phi} \\ G_{\pm} \frac{1}{\sqrt{2\pi}} e^{i(j+\frac{1}{2})\phi} \end{pmatrix}.$$

Note that H_r and $H_r^{\dagger}$ have the same expressions as differential operators although their domains could be different. If the above equations have no square-integrable solutions, then the operator H_r is self-adjoint in the domain $D(H_r)$. However, if there are equal number(s) of square integrable solutions, then the Dirac operator H_r is not self-adjoint in the domain $D(H_r)$, but can be made self-adjoint by suitable choice of domains.

The above equations can be written as

$$r\frac{du_\pm}{dr} + (\nu + i\beta \mp 2r)u_\pm - jv_\pm = 0,$$

$$r\frac{dv_\pm}{dr} + (\nu - i\beta)v_\pm - ju_\pm = 0.$$

In terms of the variable $z = -2ikr$, these equations take the form

$$z\frac{d^2v_\pm}{dz^2} + (1 + 2\nu - z)\frac{dv_\pm}{dz} - (\nu - i\beta)\, v_\pm = 0.$$

Note that $z = \pm 2r$ for $k = \pm i$, respectively.

According to the prescription by von Neumann, we have to look for square-integrable solutions of the above equations. That would give us the information about the deficiency indices. We first consider the case with $k = +i$. A square-integrable solution can be written as

$$v_+ = U(\nu - i\beta,\; 1 + 2\nu,\; z) = U(\nu - i\beta,\; 1 + 2\nu,\; 2r),$$

where $U(a, b, z)$ denotes the confluent hypergeometric function of the second kind given by

$$U(a, b, z) = \frac{\pi}{\sin \pi b}\left[\frac{M(a, b, z)}{\Gamma(1 + a - b)\Gamma(b)} - z^{1-b}\frac{M(1 + a - b, 2 - b, z)}{\Gamma(a)\Gamma(2 - b)}\right],$$

where $M(a, b, z)$ confluent hypergeometric function of the first kind. Using the above, we find that

$$u_+ = \frac{\nu - i\beta}{j}\left[U(\nu - i\beta,\; 1 + 2\nu,\; 2r) - 2rU(\nu + 1 - i\beta,\; 2 + 2\nu.\; 2r)\right],$$

Finally, we arrive at

$$F_+(r) = \left[\frac{j + \nu - i\beta}{j}U(\nu - i\beta,\; 1 + 2\nu,\; 2r)r^{\nu - \frac{1}{2}}\right.$$
$$\left. - \frac{2(\nu - i\beta)}{j}U(\nu + 1 - i\beta,\; 2 + 2\nu,\; 2r)r^{\nu + \frac{1}{2}}\right]e^{-r},$$

where we have used the relation $U'(a, b, z) = -aU(a+1, b+1, z)$. This gives the upper component of the radial part of the wave-function corresponding to $k = +i$.

In order to proceed, we need to find the condition under which F_+ is square integrable. In the limit $r \to \infty$, $U(a, b, 2r) \sim r^{-a}$ and $F_+ \to 0$, meaning that F_+ is square integrable at infinity. As $r \to 0$, $M(a, b, 2r) \to 1$

$$\int |F_+|^2 r\, dr \sim \int r^{-2\nu}\, dr$$

plus terms which are convergent. Therefore, for the range $0 < \nu < \frac{1}{2}$, F_+ is a square integrable function. Similar analysis shows that for $0 < \nu < \frac{1}{2}$, the entire radial wave-function is square integrable. Hence, we have a single square integrable solution corresponding to $k = +i$. The deficiency index n_+ of H_r is defined by the number of linearly independent and square integrable eigenfunctions corresponding to the eigenvalue $k = +i$. Our analysis shows that for gapless graphene with a charge impurity, $n_+ = 1$ when $0 < \nu < \frac{1}{2}$.

Next we consider the case when $k = -i$. Here we have $z = -2ikr = -2r$. A candidate for a square-integrable solution for v_- is given by

$$v_- = e^z U(1 + \nu + i\beta, \ 1 + 2\nu, \ -z) = e^{-2r} U(1 + \nu + i\beta, \ 1 + 2\nu, \ 2r).$$

From this, the expression for u_- is obtained as

$$u_- = \frac{\nu - i\beta}{j} e^{-2r} U(1 + \nu + i\beta, \ 1 + 2\nu, \ 2r)$$

$$- \frac{2r}{j} e^{-2r} \Big[U(1 + \nu + i\beta, \ 1 + 2\nu, \ 2r)$$

$$+ (1 + \nu + i\beta) U(2 + \nu + i\beta, \ 2 + 2\nu, \ 2r) \Big].$$

Using of the recursive relation $U(a, b, z) - aU(a+1, b, z) - U(a, b-1, z) = 0$, the expression for u_- can be simplified as

$$u_- = \frac{\nu - i\beta}{j} e^{-2r} U(1 + \nu + i\beta, \ 1 + 2\nu, \ 2r)$$

$$- \frac{2r}{j} e^{-2r} U(1 + \nu + i\beta, \ 2 + 2\nu, \ 2r),$$

so that we finally have

$$F_-(r) = e^{-r} \Big[\frac{j + \nu - i\beta}{j} r^{\nu - \frac{1}{2}} U(1 + \nu + i\beta, \ 1 + 2\nu, \ 2r)$$

$$- \frac{2}{j} r^{\nu + \frac{1}{2}} U(1 + \nu + i\beta, \ 2 + 2\nu. \ 2r) \Big]$$

This gives the upper component of the radial part of the wave-function corresponding to $k = -i$. By going through the similar procedure as before, we see that as $r \to \infty$, $F_-(r) \to 0$. In the limit when $r \to 0$, we can again show that F_- and the corresponding entire radial wave-function is square integrable when $0 < \nu < \frac{1}{2}$. Thus, for $k = -i$ also, we have a single square integrable solution when range $0 < \nu < \frac{1}{2}$. This proves that when the parameter ν satisfies the condition $0 < \nu < \frac{1}{2}$, the deficiency indices of the Dirac operator for graphene in the presence of a sub-critical external Coulomb charge impurity is given by $(n_+, n_-) = (1, 1)$.

According to the prescription of von Neumann, this result $n_+ = n_- = 1$, implies that H_r itself is not self-adjoint in the domain $D(H_r)$, but can be made self-adjoint through a suitable choice of boundary conditions. When $n_+ = n_- = 1$, the allowed boundary conditions are labelled by 1×1 unitary matrix, which is just a phase given by e^{iz}, where $z \in R$ mod 2π. The quantity z is called the self-adjoint extension parameter. In general, for each value of z we have a different quantum theory. Thus the self-adjoint extension provides a one-parameter inequivalent quantization of the graphene system with a sub-critical external Coulomb charge impurity. Note that we have also used the symbol z for the quantity $-2ikr$ which appeared in the Dirac equation. This is a minor abuse of notations but it should cause no confusion.

Having obtained the self-adjoint extension we must now find the domain in which the radial Dirac operator H_r is self-adjoint. As discussed in Chapter 4, the prescription due to von Neumann says that domain in which H_r is self-adjoint is given by $\mathcal{D}_z(H_r) = \mathcal{D}(H_r) \oplus \{e^{i\frac{z}{2}}\Psi_+ + e^{-i\frac{z}{2}}\Psi_-\}$. Physically, the domain $\mathcal{D}_z(H_r)$ provides the boundary conditions for which the radial Dirac operator for graphene is self-adjoint and we see that the boundary conditions are labelled by the parameter z.

In subsequent analysis it will be of interest to know the short distance behaviour of functions F_+ and F_-, which are given by

$$F_+ \longrightarrow \frac{\pi}{\sin\pi(1+2\nu)} \frac{\nu+j-i\beta}{j} \frac{1}{\Gamma(-\nu-i\beta)\Gamma(1+2\nu)} r^{\nu-\frac{1}{2}}$$
$$+ \left[\frac{\pi}{\sin\pi(2+2\nu)} \frac{\nu-i\beta}{j} \frac{2^{-2\nu}}{\Gamma(1+\nu-i\beta)\Gamma(-2\nu)}\right.$$
$$\left. - \frac{\pi}{\sin\pi(1+2\nu)} \frac{\nu+j-i\beta}{j} \frac{2^{-2\nu}}{\Gamma(\nu-i\beta)\Gamma(1-2\nu)}\right] r^{-\nu-\frac{1}{2}},$$

and

$$F_- \longrightarrow \frac{\pi}{\sin\pi(1+2\nu)} \frac{\nu+j-i\beta}{j} \frac{1}{\Gamma(1-\nu+i\beta)\Gamma(1+2\nu)} r^{\nu-\frac{1}{2}}$$
$$+ \left[\frac{\pi}{\sin\pi(2+2\nu)} \frac{1}{j} \frac{2^{-2\nu}}{\Gamma(1+\nu+i\beta)\Gamma(-2\nu)}\right.$$
$$\left. - \frac{\pi}{\sin\pi(1+2\nu)} \frac{\nu+j-i\beta}{j} \frac{2^{-2\nu}}{\Gamma(1+\nu+i\beta)\Gamma(1-2\nu)}\right] r^{-\nu-\frac{1}{2}}.$$

This concludes our discussion of the allowed boundary conditions for the radial Dirac operator for the gapless graphene.

5.3.2 Scattering Matrix for Gapless Graphene with Coulomb Charge

We now turn to problem of finding the scattering state solutions of the eigenvalue problem for the parameter range $0 < \nu < \frac{1}{2}$. For this, the appropriate solution for the function v is given by

$$v = A_1 M\left(\nu - i\beta,\ 1 + 2\nu,\ z\right) + A_2 z^{-2\nu} M\left(-\nu - i\beta,\ 1 - 2\nu,\ z\right),$$

where $z = -2ikr$ and A_1 and A_2 are constants which will be determined later.

With this form of v, the function u can be obtained as

$$
\begin{aligned}
ju = {} & A_1 z \frac{d}{dz} M\left(\nu - i\beta,\ 1 + 2\nu,\ z\right) \\
& + A_2 z (-2\nu) z^{-2\nu - 1} M\left(-\nu - i\beta,\ 1 - 2\nu,\ z\right) \\
& + A_2 z^{-2\nu + 1} \frac{d}{dz} M\left(-\nu - i\beta,\ 1 - 2\nu,\ z\right) \\
& + A_1 (\nu - i\beta) M\left(\nu - i\beta,\ 1 + 2\nu,\ z\right) \\
& + A_2 (\nu - i\beta) z^{-2\nu} M\left(-\nu - i\beta,\ 1 - 2\nu,\ z\right).
\end{aligned}
$$

Using the relation

$$aM(a, b, z) + z M'(a, b, z) = a M(a + 1, b, z),$$

where prime denotes a derivation with respect to z, we get

$$
\begin{aligned}
ju = {} & A_1 (\nu - i\beta) M\left(1 + \nu - i\beta,\ 1 + 2\nu,\ z\right) \\
& + A_2 z^{-2\nu} (-\nu - i\beta) M\left(1 - \nu - i\beta,\ 1 - 2\nu,\ z\right).
\end{aligned}
$$

This leads to

$$
\begin{aligned}
F(r) = r^{\nu - \frac{1}{2}} e^{ikr} \Big[& A_1 \frac{\nu - i\beta}{j} M\left(1 + \nu - i\beta,\ 1 + 2\nu,\ z\right) \\
& + A_2 z^{-2\nu} \frac{-\nu - i\beta}{j} M\left(1 - \nu - i\beta,\ 1 - 2\nu,\ z\right) \\
& + A_1 M\left(\nu - i\beta,\ 1 + 2\nu,\ z\right) \\
& + A_2 z^{-2\nu} M\left(-\nu - i\beta,\ 1 - 2\nu,\ z\right) \Big],
\end{aligned}
$$

for the radial part of the first component of the wave-function Ψ of the Dirac equation.

Having obtained the solution of the Dirac equation, our task is now to impose the boundary conditions. In this case, the boundary conditions are encoded in the domain of self-adjointness of the corresponding Dirac operator, given by $\mathcal{D}_z(H_r)$. This followed from the prescription of von Neumann. How do we fit the physical wave-function to this boundary condition? For that we adopt the following strategy. First we find the behaviour of a generic element of the domain $\mathcal{D}_z(H_r)$ in the limit $r \to 0$. Next we find the behaviour of the solution of the Dirac equation in the same limit. Now we need to ensure that the solution of the Dirac equation lies in the domain of self=adjointness. Within our framework this is ensured by demanding that in these two expressions, the coefficients of the same powers of the variable r matches in the limit $r \to 0$. This is equivalent to imposing the boundary condition that will ensure self-adjointness of the Dirac operator and the unitarity of the quantum time evolution.

To see how this works, first consider the physical wavefunction. In the limit $r \to 0$

$$F(r) \longrightarrow A_1 \left(1 + \frac{\nu - i\beta}{j}\right) r^{\nu - \frac{1}{2}} + A_2(-2ik)^{-2\nu} \left(1 + \frac{-\nu - i\beta}{j}\right) r^{-\nu - \frac{1}{2}}.$$

In the same limit a typical element of the domain $\mathcal{D}_z(H_r)$ is given by

$$\Psi(r, \phi) = \lambda \left(e^{i\frac{z}{2}} \Psi_+ + e^{-i\frac{z}{2}} \Psi_-\right),$$

where λ is a constant. Comparing the upper component of the physical solution to that of an element of $\mathcal{D}_z(H_r)$ in the limit $r \to 0$, we get

$$A_1 = \lambda \frac{\pi}{\sin \pi(1 + 2\nu)} \left(\frac{e^{i\frac{z}{2}}}{\Gamma(-\nu - i\beta)\Gamma(1 + 2\nu)} + \frac{e^{-i\frac{z}{2}}}{\Gamma(1 - \nu + i\beta)\Gamma(1 + 2\nu)}\right),$$

$$A_2(-2ik)^{-2\nu}(j - \nu - i\beta) = -\lambda \frac{\pi}{\sin \pi(1 + 2\nu)} \left[(\nu - i\beta) \frac{2^{-2\nu} e^{i\frac{z}{2}}}{\Gamma(1 + \nu - i\beta)\Gamma(-2\nu)}\right.$$

$$+ (\nu + j - i\beta) \frac{2^{-2\nu} e^{i\frac{z}{2}}}{\Gamma(\nu - i\beta)\Gamma(1 - 2\nu)}$$

$$+ \frac{2^{-2\nu} e^{-i\frac{z}{2}}}{\Gamma(1 + \nu + i\beta)\Gamma(-2\nu)}$$

$$\left. + (\nu + j - i\beta) \frac{2^{-2\nu} e^{-i\frac{z}{2}}}{\Gamma(1 + \nu + i\beta)\Gamma(1 - 2\nu)}\right].$$

This procedure ensures that the physical wavefunction satisfies the correct boundary condition.

In order to find the phase shifts and the scattering matrix, we have to look at the the asymptotic behaviour of the wave-function Ψ, or equivalently, of the function $F(r)$. To do this, we note that the behaviour of

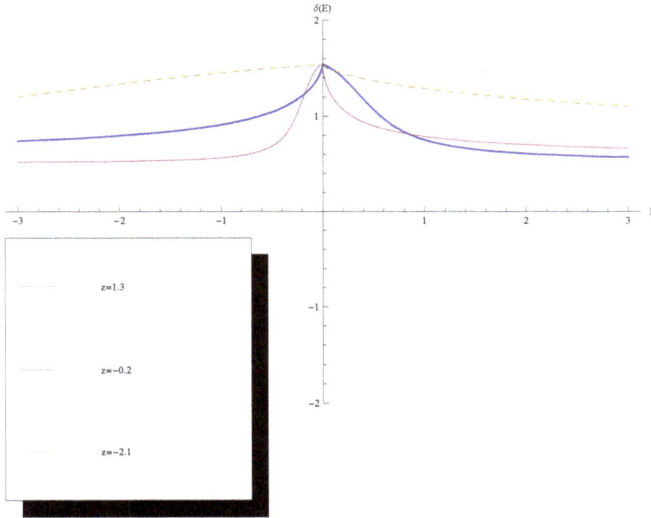

Fig. 5.2   Phase shifts in the gapless graphene for values of the self-adjoint extension parameter $z = 1.3$, -0.2, -2.1 and $j = -\frac{1}{2}$, $\beta = 0.38$ and $\nu = 0.32$.

the confluent hypergeometric function $M(a, b, z)$ in the asymptotic region $\mathrm{Re}(z) = 0$ and $\mathrm{Im}(z) \to +\infty$ is given by

$$M(a, b, z) \longrightarrow \frac{\Gamma(b)}{\Gamma(a)} e^z z^{a-b} \left[1 + O(|z|^{-1})\right] + \frac{\Gamma(b)}{\Gamma(b-a)} z^{-a} e^{\pm i\pi a} \left[1 + O(|z|^{-1})\right].$$

Due to the fact that we are dealing with the problem where $\mathrm{Re}(z) = 0$, both leading terms in the above asymptotic expansion of M give comparable contribution of the same order. Hence both of them have to be considered. Thus the behaviour of $F(r)$ when $r \to \infty$ is given by

$$
\begin{aligned}
F(r) \longrightarrow\ & (-2ik)^{-\nu}(-i)^{-i\beta} \left(A_1 \frac{\nu - i\beta}{j} \frac{\Gamma(1 + 2\nu)}{\Gamma(1 + \nu - i\beta)} \right. \\
& \left. + A_2 \frac{-\nu - i\beta}{j} \frac{\Gamma(1 - 2\nu)}{\Gamma(1 - \nu - i\beta)} \right) \frac{e^{-i(kr + \beta \ln 2kr)}}{\sqrt{r}} \\
& + (-2ik)^{-\nu}(-i)^{i\beta} \left(A_1 \frac{\Gamma(1 + 2\nu)}{\Gamma(1 + \nu + i\beta)} e^{-i\pi(\nu - i\beta)} \right. \\
& \left. + A_2 \frac{\Gamma(1 - 2\nu)}{\Gamma(1 - \nu + i\beta)} e^{i\pi(\nu + i\beta)} \right) \frac{e^{i(kr + \beta \ln 2kr)}}{\sqrt{r}}.
\end{aligned}
$$

The scattering matrix and the corresponding phase shift can be written

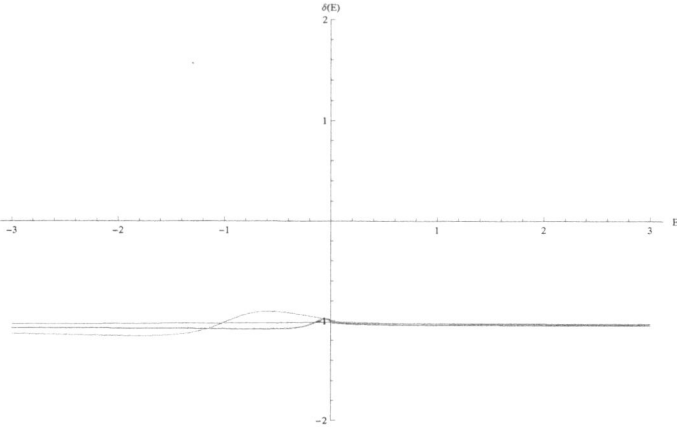

Fig. 5.3   Phase shifts in the gapless graphene for values of the self-adjoint extension parameter $z = 1.3$, -0.2, -2.1 and with $j = \frac{1}{2}$, $\beta = -0.49$ and $\nu = 0.1$.

as

$$S(k) \equiv e^{2i\delta(k)} = e^{-4\pi\beta} \frac{\frac{A_1}{A_2} \frac{\Gamma(1+2\nu)}{\Gamma(1+\nu+i\beta)} e^{-i\pi\nu} + \frac{\Gamma(1-2\nu)}{\Gamma(1-\nu+i\beta)} e^{i\pi\nu}}{\frac{A_1}{A_2} \frac{\nu-i\beta}{j} \frac{\Gamma(1+2\nu)}{\Gamma(1+\nu-i\beta)} + \frac{-\nu-i\beta}{j} \frac{\Gamma(1-2\nu)}{\Gamma(1-\nu-i\beta)}},$$

where the ratio A_1/A_2 is given by

$$\frac{A_1}{A_2} = -(-ik)^{-2\nu} \frac{\Gamma(1-2\nu)}{\Gamma(1+2\nu)} \frac{\frac{e^{i\frac{z}{2}}}{\Gamma(-\nu-i\beta)} + \frac{e^{-i\frac{z}{2}}}{\Gamma(1-\nu+i\beta)}}{\frac{e^{i\frac{z}{2}}}{\Gamma(\nu-i\beta)} + \frac{e^{-i\frac{z}{2}}}{\Gamma(1+\nu+i\beta)}}.$$

The S matrix has complete information about the spectrum of the theory. In particular, the poles of the S matrix in the positive imaginary axis corresponds to the bound states. From the expression of the S matrix we see that it does not have any poles in the positive imaginary value of the energy. Thus the subcritical regime of a gapless graphene does not have any bound states even in the presence of self-adjoint extension. This may look surprising since the self-adjoint extension introduces a scale in the problem, which could have generated a bound state. However, the absence of the bound state for gapless graphene is in qualitative agreement with the Klein tunneling principle, which precludes the formation of such bound states.

The above analysis holds only when the system parameters are such that $0 < \nu < \frac{1}{2}$, where $\nu = \sqrt{j^2 - \beta^2}$. When ν belongs to this range, then for every value of the self-adjoint extension parameter $z \in R \mod 2\pi$, we have a different quantum description of the system. All the physical observables of the system, would depend on the choice of z. To see how self-adjoint extension parameter z affects the physics, consider the scattering phase shift $\delta(k)$, $k = -E$, as a function of energy.

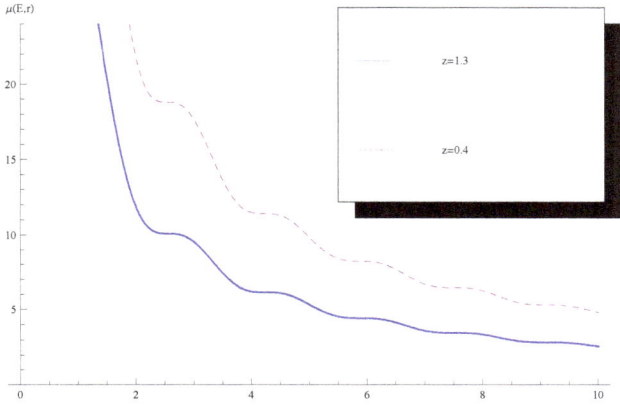

Fig. 5.4 Local density of states in the gapless graphene as a function of distance from impurity, obtained by summing the contributions coming from angular momentum channels $j = \frac{1}{2}$ and $j = -\frac{1}{2}$, where both contributions correspond to energy $E = 2$. The self-adjoint extension parameter has been chosen as $z = 1.3$ and $z = 0.4$. Other system parameters are $\beta = 0.4$, $\nu = 0.3$ and $N_c = 4$.

Fig. 5.2 shows the energy dependence of $\delta(k)$ for an attractive Coulomb potential. The phase shift not only depends on the energy, but it also has different behaviour for different choices of the self-adjoint extension parameter z. For the case of the repulsive Coulomb potential as shown in Fig. 5.3, the phase shift is almost independent of the energy and the dependence on z is also less prominent. It is possible to relate the phase shift to polarization charge using the Friedel sum rule. Our analysis indicates that the polarization charge would explicitly depend on the choice of the boundary condition through their dependence on z.

The local density of states (LDOS) is defined as

$$\mu(E, r) = \frac{N_c}{\pi \hbar v_F} \sum_j |\Psi(r, \phi; E, j)|^2,$$

where N_c is the combined valley and spin degeneracy in graphene, which has the value $N_c = 4$. The LDOS has been calculated numerically and plotted

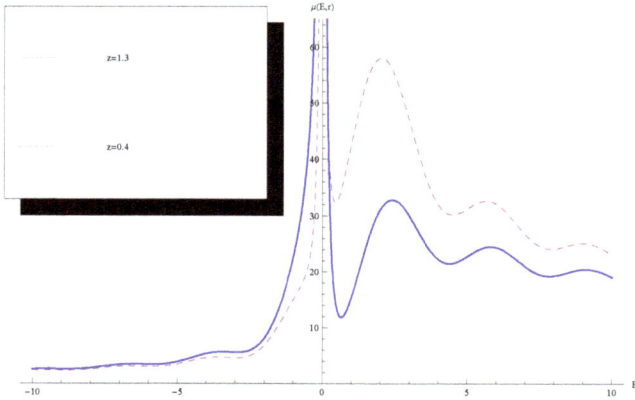

Fig. 5.5   Energy dependence of the local density of states in the gapless graphene at the distance $r = 1$ from the impurity, obtained by summing the contributions coming from the angular momentum channels $j = \frac{1}{2}$ and $j = -\frac{1}{2}$. The self-adjoint extension parameter has been chosen as $z = 1.3$ and $z = 0.4$. Other system parameters have values $\beta = 0.4$, $\nu = 0.3$ and $N_c = 4$.

in Fig. 5.4 and Fig. 5.5. It is clear from the plots that the behaviour of the LDOS again explicitly depend on the choice of z.

We have therefore seen that the phase shifts, the S-matrix, the LDOS and the associated physical observables all depend on the choice of the self-adjoint extension parameter z. At this stage we might ask about the physical reasons for such a dependence. We know that z encodes all allowed boundary conditions which are consistent with a unitary time evolution. But in the theoretical description that we have presented, why is there such an ambiguity? One way to understand this is to recall that the Dirac description of graphene is valid only in the long wavelength limit. When a Coulomb charge impurity is placed on a graphene sheet, it generates a Coulomb potential. In addition, due to the local interaction of the impurity with the graphene lattice, there could be many other short range interactions whose details are unknown to us. Even if we knew some of them, it would not have been possible to incorporate those short distance interactions in the Dirac equation, which is valid only in the long wavelength limit. It is at this stage that the method of self-adjoint extension due to von Neumann offers a way out of this apparent impasse. This method tells us that in this case, the combined effect of all those short range interactions can be incorporated in a single parameter z. The theory is not able to specify the exact value of z for a given system. That has to be determined empirically. The method of self-adjoint extension is thus a powerful and elegant tool to describe effect of boundary conditions.

5.3.3 Gapless Graphene with Supercritical Coulomb Charge

In this Section we shall discuss the case where the strength of the external Coulomb charge impurity exceeds the critical value. In this case, there is a breakdown of the quantum vacuum and the energy of the system governed by the Dirac equation becomes unbounded from below. Such a situation in quantum mechanics was first envisaged by Landau, who famously described it as *fall to the centre*. In a real graphene sample the lattice provides an ultraviolet cutoff which prevents the energy to go all the way to negative infinity. Nevertheless, graphene provides a unique opportunity to study this strong QED effect and it is instructive to analyze the Dirac equation in this context.

In order to understand the critical charge, recall that the Dirac wavefunction was written as

$$\Psi(r, \phi) = \begin{pmatrix} F(r) \, \frac{1}{\sqrt{2\pi}} e^{i(j-\frac{1}{2})\phi} \\ G(r) \, \frac{1}{\sqrt{2\pi}} e^{i(j+\frac{1}{2})\phi} \end{pmatrix},$$

with

$$F(r) = e^{ikr} r^{\nu-\frac{1}{2}} (u(r) + v(r)),$$
$$G(r) = e^{ikr} r^{\nu-\frac{1}{2}} (u(r) - v(r)),$$

Note that $\nu = \sqrt{j^2 - \beta^2}$ and $\beta = \frac{Ze^2}{\kappa \hbar v_F}$, which measures the strength of the electromagnetic interaction due to the external Coulomb charge impurity. The minimum value of j in planar graphene is $\frac{1}{2}$. Thus, if β becomes more that $\frac{1}{2}$, then ν becomes imaginary in the $j = \frac{1}{2}$ channel. In that case the factor $r^{\nu-\frac{1}{2}}$ in the wavefunction is no longer real and it becomes wildly oscillatory as $r \to 0$. This indicates the breakdown of the quantum vacuum and signals the onset of the fall to the centre phenomenon. This is nothing but a strong nonperturbative QED effect. The critical value of β is denoted by β_c. For planar graphene $\beta_c = \frac{1}{2}$. If $\beta > \beta_c$, the external Coulomb charge impurity is said to be supercritical. As discussed at the beginning of this Chapter, the coupling $\beta \sim 1$ for $Ze \sim 1$. It is thus relatively easy to reach the strong nonperturbative regime in graphene.

We would now like to solve the Dirac equation with supercritical charge impurity. For that, define a new function η as

$$v = \rho^{-(\frac{1}{2}+\nu)} e^{-ik\rho} \eta.$$

The differential equation for v that was derived before can now be written as

$$\frac{d^2\eta}{dz^2} + \left[-\frac{1}{4} + \frac{\frac{1}{2} + i\beta}{z} + \frac{(\frac{1}{4} - \nu^2)}{z^2} \right] \eta = 0,$$

This has the form of Whittaker's equation.

When $\beta > \beta_c$, $\nu = \pm i\mu$, $\mu = \sqrt{\beta^2 - m^2}$ and the above equation can be written as

$$\frac{d^2\eta}{dz^2} + \left[-\frac{1}{4} + \frac{\frac{1}{2} + i\beta}{z} + \frac{(\frac{1}{4} + \mu^2)}{z^2} \right] \eta = 0.$$

The general solution of this is given by

$$\eta(z) = A_1 e^{i\theta} M_{\frac{1}{2} + i\beta, i\mu}(z) + A_2 e^{-i\theta} M_{\frac{1}{2} + i\beta, -i\mu}(z),$$

where A_1 and A_2 are constants and

$$M_{\frac{1}{2} + i\beta, \pm i\mu} = e^{-\frac{z}{2}} z^{\frac{1}{2} \pm i\mu} M(\pm i\mu - i\beta, 1 \pm 2i\mu, z),$$

with M on the right hand side denoting the confluent hypergeometric function.

We shall now construct solutions which are square integrable at infinity. Let us first note that for the functions $F(r)$ and $G(r)$. The measure of integration is given by rdr. Consider the contribution of the function η to the quantity $\int |F|^2 rdr$ at asymptotic infinity. Using the asymptotic formula for the confluent hypergeometric function, we find that as $|z| = |-2ikr| \to \infty$,

$$M_{\frac{1}{2} + i\beta, \pm i\mu}(z) \longrightarrow e^{ikr} e^{\pi(\pm\mu - \beta)} (-2ikr)^{\frac{1}{2} + i\beta} \frac{\Gamma(1 \pm 2i\mu)}{\Gamma(1 \pm i\mu + i\beta)}$$

$$+ e^{-ikr} (-2ikr)^{-\frac{1}{2} - i\beta} \frac{\Gamma(1 \pm 2i\mu)}{\Gamma(\pm i\mu - i\beta)}.$$

As $r \to \infty$, the contribution of the first term on the right hand side of the above equation to $|F|^2$ behaves as $\sim \frac{1}{r}$ while that from the second term behaves as $\sim \frac{1}{r^3}$. Hence, the first term provides a divergent contribution to $\int |F|^2 rdr$ while the contribution of the second term is convergent. We now consider the part of the solution η arising from the first term in the right hand side, which leads to the divergence of the norm of F. This is denoted by η_{div} and is given by

$$\eta_{\text{div}} \to e^{ikr} (-2ikr)^{\frac{1}{2} + i\beta} e^{-\pi\beta} \left[A_1 e^{i\theta} e^{\pi\mu} \frac{\Gamma(1 + 2i\mu)}{\Gamma(1 + i\mu + i\beta)} \right.$$

$$\left. + A_2 e^{-i\theta} e^{-\pi\mu} \frac{\Gamma(1 - 2i\mu)}{\Gamma(1 - i\mu + i\beta)} \right].$$

We now choose $A_2 = A_1 e^{2\pi\mu} \frac{\Gamma(1-i\mu+i\beta)}{\Gamma(1-i\mu-i\beta)}$ and denote $\frac{\Gamma(1+2i\mu)}{\Gamma(1+i\mu-i\beta)} = \xi e^{i\gamma}$. Therefore, as $r \to \infty$, we get

$$\eta_{\text{div}} \to\to 2A_1 e^{ik\rho}(-2ik\rho)^{\frac{1}{2}+i\beta} e^{\pi(\mu-\beta)} \xi \cos(\theta + \gamma).$$

The quantity η_{div} can be made to vanish if

$$\cos(\theta + \gamma) = 0 \quad \text{or} \quad \theta = -\gamma + \left(n + \frac{1}{2}\right)\pi, \quad n \in Z.$$

This condition leads to a quantization of the parameter θ. The quantization of θ ensures that as $\rho \to \infty$, $\int |F|^2 r dr$ will remain finite. Knowing η, the components u and v of the solution to the Dirac equation can be obtained. They are square-integrable and represent the bound state solution.

We now proceed to obtain the quantized energy levels. For that purpose, consider the short distance limit of the general solution given by η. We write $\frac{\Gamma(1-i\mu+i\beta)}{\Gamma(1-i\mu-i\beta)} = Ce^{i2\delta}$. Then, as $z \to 0$,

$$\eta \sim \sqrt{z}\left[(1 + C)\cos(\theta - \delta + \mu \ln z) + (1 - C)\sin(\theta - \delta + \mu \ln z)\right].$$

The short distance behaviour of the system can also be inferred by looking at the corresponding indical equation. In this case, the roots of the indical equation are given by $\frac{1}{2} \pm i\mu$. Using these two roots, the wavefunction at short distance can be obtained as

$$\eta \sim \sqrt{-2ir}\left[e^{iB}e^{i\mu \ln(-2ir)} + Ce^{-iB}e^{-i\mu \ln(-2ir)}\right]$$

$$\sim \sqrt{-2ir}\left[(1 + C)\cos(B + \mu \ln(-2ir))\right.$$

$$\left. + i(1 - C)\sin(B + \mu \ln(-2ir))\right],$$

where B is a real constant. The short distance behaviour of the wavefunction obtained from the two approaches above must match. This is possible provided

$$\theta + \delta + \mu \ln k = B.$$

Thus we obtain the quantized energy eigenvalues as

$$E_n = -\frac{kn}{\hbar v_F} = -\frac{e^{-(n+\frac{1}{2})\frac{\pi}{\mu}+A}}{\hbar v_F}, \quad n \in Z,$$

where $A = \frac{(B-\delta)}{\mu}$ is a constant. The quantity δ depends on the system parameters μ and β. B on the other hand is a real constant which is not fixed by the theoretical analysis. This leads to a one parameter family of inequivalent spectra in the supercritical regime. As mentioned before, B encodes the effect of short distance physics, which are expected to be important as we approach length scales of the order of the lattice spacings. The analysis suggests that it is not important to know the the details of the short distance interactions and their effect on the spectrum appears through a single parameter B, which should be determined empirically. The ratio of the various energy levels are independent of the parameter B. It may be noted that the above quantization has been carried out with only the function $v(r)$. The function $u(r)$ can be determined from the Dirac equation and it can be shown that it does not change the quantization condition.

We have thus found an infinite number of bound states in graphene containing a impurity, when the effective charge exceeds the critical value. These states are square-integrable at infinity and have a rapidly oscillatory behaviour in the short distance limit, which is a feature of the fall to the centre. Our analysis is fully quantum and nonperturbative, valid for any value of $\beta > \beta_c$. As expected, the spectrum we get is unbounded from below, which is an important feature of the supercritical region. In reality the lattice spacing in graphene will provide a natural cut-off due to which the energy would not diverge. In addition, we find an accumulation point at zero energy.

Another interesting feature of graphene is that the screening in the supercritical region drives the effective charge to the critical value. In our formulation this is achieved when $\mu \to 0$ from the supercritical region. Within the scope of the quantum mechanics discussed here, this effect can be viewed as a renormalization group flow arising from the short distance effects. In the short distance limit, the eigenvalue equation assumes the form

$$\frac{d^2\eta}{dr^2} + \frac{(\frac{1}{4} + \mu^2)}{r^2}\eta = \lambda\eta,$$

with the eigenvalue $\lambda = 0$. In the supercritical region, the spectrum is unbounded from below. In order to regulate this divergence, we introduce a short distance cutoff at $r = a$ and impose the boundary condition that the wave-function vanishes below the cutoff. With such a boundary condition, the eigenvalues can be calculated analytically for small values of μ, leading to a finite bound state spectrum for λ given by

$$\lambda_n = -e^{-\frac{2n\pi}{\mu}}\left[\frac{2}{ae^\sigma}\right],$$

where σ is the Euler's constant and $n = 1, 2, ..., \infty$. Note that for any finite value of the cutoff, as $n \to \infty$, $\lambda_n \to 0$ and the zero eigenvalue depicts an accumulation point for this system. In this limit we recover the short distance form of the original eigenvalue problem. The eigenvalues λ_n also explicitly depend on the cutoff and they diverge as the cutoff is removed. This indicates that the breakdown of the 2D Dirac description of graphene at short distances. In the spirit of renormalization group analysis, we now make the coupling μ a function of the cutoff, or take $\mu = \mu(a)$. In order to find the dependence of μ on the cutoff, we demand that the zero eigenvalue remains unchanged as the cutoff is removed. This leads to a beta function for the coupling μ given by

$$\tilde{\beta}(\mu) = -a\frac{d\mu}{da} \approx -\mu^2.$$

This beta function has an ultraviolet stable fixed point at $\mu = 0$. In other words, the effective charge is driven to its critical value as the cutoff is removed. The same result can be obtained from the analysis of nonlinear effects on screening.

5.4 Gapped Graphene with Coulomb Charge Impurity

Pristine graphene is a gapless material. If the sublattice symmetry in graphene is broken due to some external effect, a gap can open up. A gapped graphene in principle is more suited for various technological applications. It is therefore of interest to study the properties of gapped graphene. In the long wavelength limit, the standard way to introduce a gap is by incorporating a mass term in the Dirac equation for graphene. We shall follow this approach. Our focus will be to study the properties of the system in the presence of an external charge impurity. We shall see that there are important differences in the spectrum compared to the gapless case.

For a gapped monolayer planar graphene, the Dirac equation can be written as

$$H\Psi = E\Psi,$$

where the Hamiltonian is given by

$$H = -i\hbar v_F(\sigma_1\partial_x + \sigma_2\partial_y), +m\sigma_3 - \frac{Ze^2}{\kappa r}$$

where m is the mass of the quasiparticle indicating the gapped case and v_F is the Fermi velocity in graphene. As in the gapless case, the external

Coulomb charge provides the $\frac{Ze^2}{\kappa r}$ term to the interaction, where κ is the effective dielectric constant. The wavefunction has the form

$$\Psi(r,\phi) = \begin{pmatrix} \psi_1(r) \ \frac{1}{\sqrt{2\pi}} e^{i(j-\frac{1}{2})\phi} \\ i\psi_2(r) \ \frac{1}{\sqrt{2\pi}} e^{i(j+\frac{1}{2})\phi} \end{pmatrix}.$$

In the above equations r and ϕ are the radial variable and the angle in the $x-y$ plane, respectively. Consider the ansatz

$$\psi_1(r) = \sqrt{m+E} \ e^{-\frac{\rho}{2}} \rho^{\nu-\frac{1}{2}} (P(\rho) + Q(\rho)),$$

$$\psi_2(r) = \sqrt{m-E} \ e^{-\frac{\rho}{2}} \rho^{\nu-\frac{1}{2}} (P(\rho) - Q(\rho)),$$

where $\rho = 2\gamma r = 2\sqrt{m^2 - E^2} \ r$ and $\nu = \sqrt{j^2 - \beta^2}$, j being a half integer. As in the gapless case, $\beta = \frac{Ze^2}{\kappa \hbar v_F}$, which is a measure of the strength of the external Coulomb charge impurity. In what follows we set $\hbar = v_F = 1$ by suitable choice of units.

Using this ansatz, the Dirac equation gives

$$H_\rho \begin{pmatrix} P \\ Q \end{pmatrix} = \begin{pmatrix} \rho\frac{d}{d\rho} + \nu - \frac{\beta E}{\gamma} & -j + \frac{m\beta}{\gamma} \\ -j - \frac{m\beta}{\gamma} & \rho\frac{d}{d\rho} + \nu - \rho + \frac{\beta E}{\gamma} \end{pmatrix} \begin{pmatrix} P \\ Q \end{pmatrix} = 0,$$

where H_ρ defined above denotes the radial Dirac operator. The above equations can be decoupled to give

$$\rho\frac{d^2 P}{d\rho^2} + (1 + 2\nu - \rho)\frac{dP}{d\rho} - \left(\nu - \frac{\beta E}{\gamma}\right) P = 0,$$

$$\rho\frac{d^2 Q}{d\rho^2} + (1 + 2\nu - \rho)\frac{dQ}{d\rho} - \left(1 + \nu - \frac{\beta E}{\gamma}\right) Q = 0.$$

Thus we see that the functions P and Q satisfy the confluent hypergeometric equation. This equation has two linearly independent solutions, one of which is regular at the origin, which we denote by M. The other solution is regular at infinity, which is denoted by U. If the boundary conditions were so chosen that the solutions are regular at the origin, then the wavefunctions are obtained as

$$\psi_1(\rho) = \sqrt{m+E} e^{-\frac{\rho}{2}} \rho^{\nu-\frac{1}{2}} [(j + \frac{m\beta}{\nu}) M(\nu - \frac{\beta E}{\gamma}, 1 + 2\nu, \rho)$$

$$+ (\nu - \frac{\beta E}{\gamma}) M(1 + \nu - \frac{\beta E}{\gamma}, 1 + 2\nu, \rho)]$$

$$\psi_2(\rho) = \sqrt{m-E} e^{-\frac{\rho}{2}} \rho^{\nu-\frac{1}{2}} [(j + \frac{m\beta}{\nu}) M(\nu - \frac{\beta E}{\gamma}, 1 + 2\nu, \rho)$$

$$- (\nu - \frac{\beta E}{\gamma}) M(1 + \nu - \frac{\beta E}{\gamma}, 1 + 2\nu, \rho)].$$

Using the requirement that the wavefunction is square integrable, we get the bound state spectrum as

$$E_{p,j} = \frac{m \; \mathrm{sgn}(\beta)}{\sqrt{1 + \frac{\alpha^2}{(p+\nu)^2}}},$$

with $p = 0, 1, 2, ...$, for $j > 0$ and $p = 1, 2, 3, ...$, for $j < 0$.

Unlike the gapless case, the Dirac equation for graphene admits bound states in the presence of an external Coulomb charge impurity. Next we shall see that more general boundary conditions are possible which are consistent with all the requirements of quantum mechanics, leading to a different spectrum for the same Dirac operator for gapped graphene.

5.4.1 Boundary Conditions for Gapped Graphene with a Charge Impurity

The external charge impurity in gapped graphene provides a Coulomb interaction. In addition, due to reasons discussed before, it can generate other short range interactions which cannot be directly incorporated in the Dirac equation, which is valid only in the long wavelength limit. An average effect of such short range interactions can be taken into account through boundary conditions. For that, we need to know what are the all allowed class of boundary conditions for this problem. As before, we shall use the method of self-adjoint extension due to von Neumann to analyze this issue.

In order to find the deficiency indices $n_\pm$ for H_ρ and subsequently for H, we need to solve equations

$$H^\dagger \Psi_\pm = \pm i \Psi_\pm,$$

where H^* has the same differential expression as for H, although their domains could be different. $\Psi_\pm$ are two-component spinors of the form

$$\Psi_\pm = \begin{pmatrix} \psi_{1\pm}(r) \; \frac{1}{\sqrt{2\pi}} e^{i(j-\frac{1}{2})\phi} \\ i\psi_{2\pm}(r) \; \frac{1}{\sqrt{2\pi}} e^{i(j+\frac{1}{2})\phi} \end{pmatrix},$$

where

$$\psi_{1\pm} = \sqrt{m \pm ie} \; e^{-\frac{\rho}{2}} \rho^{\nu-\frac{1}{2}} (P_\pm + Q_\pm)(\rho),$$
$$\psi_{2\pm} = \sqrt{m \mp ie} \; e^{-\frac{\rho}{2}} \rho^{\nu-\frac{1}{2}} (P_\pm - Q_\pm)(\rho).$$

The functions $\psi_{1\pm}$ and $\psi_{2\pm}$ have to be square integrable in R^+ with a measure $\rho d\rho$. To get any further information one has to solve for $P_\pm$ and $Q_\pm$ from the equations

$$\rho \frac{dP_\pm}{d\rho} + (\nu - \frac{i\beta}{\gamma_\pm}) P_\pm + \left(\frac{m\beta}{\gamma_\pm} - j \right) Q_\pm = 0,$$

$$\rho\frac{dQ_\pm}{d\rho} + (\nu - \rho + \frac{i\beta}{\gamma_\pm})Q_\pm - \left(j + \frac{m\beta}{\gamma_\pm}\right)P_\pm = 0,$$

where $\gamma_\pm = \sqrt{m^2+1}$ and $\rho = 2\gamma_\pm r$. These equations can also be decoupled to give

$$\rho\frac{d^2P_\pm}{d\rho^2} + (1+2\nu-\rho)\frac{dP_\pm}{d\rho} - \left(\nu - \frac{i\beta}{\gamma_\pm}\right)P_\pm = 0$$

$$\rho\frac{d^2Q_\pm}{d\rho^2} + (1+2\nu-\rho)\frac{dQ_\pm}{d\rho} - \left(1+\nu - \frac{i\beta}{\gamma_\pm}\right)Q_\pm = 0,$$

Consider the solution determined by the positive sign in the above expression, which is square integrable at infinity. That is given by

$$P_+ = U\left(\nu - \frac{i\beta}{\gamma_+}, 1+2\nu, \rho\right),$$

where U denotes a confluent hypergeometric function. Q_+ can be obtained from

$$\left(j - \frac{m\beta}{\gamma_+}\right)Q_+ = \rho\frac{d}{d\rho}U\left(\nu - \frac{i\beta}{\gamma_+}, 1+2\nu, \rho\right) + \left(\nu - \frac{i\beta}{\gamma_+}\right)U\left(\nu - \frac{i\beta}{\gamma_+}, 1+2\nu, \rho\right).$$

Using the relation $zU'(a,b,z) + aU(a,b,z) = a(1+a-b)U(a+1,b,z)$, where the prime denotes the derivative with respect to z, we get

$$\left(j - \frac{m\alpha}{\gamma_+}\right)Q_+ = \left(\nu - \frac{i\alpha}{\gamma_+}\right)\left(-\nu - \frac{i\alpha}{\gamma_+}\right)U\left(1+\nu - \frac{i\alpha}{\gamma_+}, 1+2\nu, \rho\right).$$

In the limit $\rho \longrightarrow 0$ the functions P_+ and Q_+ behave as

$$P_+ \longrightarrow a(A_+ - B_+\rho^{-2\nu}),$$
$$Q_+ \longrightarrow a(C_+ - D_+\rho^{-2\nu}),$$

where $a = \frac{\pi}{\sin\pi(1+2\nu)}$ and

$$A_+ = \frac{1}{\Gamma(-\nu - \frac{i\beta}{\gamma_+})\Gamma(1+2\nu)} \qquad B_+ = \frac{1}{\Gamma(\nu - \frac{i\beta}{\gamma_+})\Gamma(1-2\nu)}$$

$$C_+ = \frac{(\nu - \frac{i\beta}{\gamma_+})(-\nu - \frac{i\beta}{\gamma_+})}{(j - \frac{m\beta}{\gamma_+})}\frac{1}{\Gamma(1-\nu - \frac{i\beta}{\gamma_+})\Gamma(1+2\nu)}$$

$$D_+ = \frac{(\nu - \frac{i\beta}{\gamma_+})(-\nu - \frac{i\beta}{\gamma_+})}{(j - \frac{m\beta}{\gamma_+})}\frac{1}{\Gamma(1+\nu - \frac{i\beta}{\gamma_+})\Gamma(1-2\nu)}$$

are constants depending on the system parameters. With these expressions it is easily seen that ψ_{1+} and ψ_{2+} are square integrable everywhere provided $0 < \nu < \frac{1}{2}$. Thus $n_+ = 1$ for the parameter range $0 < \nu < \frac{1}{2}$.

Using a similar analysis, we obtain

$$P_- = U\left(\nu + \frac{i\beta}{\gamma_+}, 1 + 2\nu, \rho\right),$$

$$\left(j - \frac{m\beta}{\gamma_+}\right)Q_- = \left(\nu + \frac{i\beta}{\gamma_+}\right)\left(-\nu + \frac{i\beta}{\gamma_+}\right)U\left(1 + \nu + \frac{i\beta}{\gamma_+}, 1 + 2\nu, \rho\right).$$

In the limit as $\rho \longrightarrow 0$, the functions P_- and Q_- behave as

$$P_- \longrightarrow a(A_- - B_-\rho^{-2\nu}),$$
$$Q_- \longrightarrow a(C_- - D_-\rho^{-2\nu}),$$

where $a = \frac{\pi}{\sin\pi(1+2\nu)}$ and

$$A_- = \frac{1}{\Gamma(-\nu + \frac{i\beta}{\gamma_+})\Gamma(1 + 2\nu)} \qquad B_- = \frac{1}{\Gamma(\nu + \frac{i\beta}{\gamma_+})\Gamma(1 - 2\nu)}$$

$$C_- = \frac{(\nu + \frac{i\beta}{\gamma_+})(-\nu + \frac{i\beta}{\gamma_+})}{(j - \frac{m\beta}{\gamma_+})} \frac{1}{\Gamma(1 - \nu + \frac{i\beta}{\gamma_+})\Gamma(1 + 2\nu)}$$

$$D_- = \frac{(\nu + \frac{i\beta}{\gamma_+})(-\nu + \frac{i\beta}{\gamma_+})}{(j - \frac{m\beta}{\gamma_+})} \frac{1}{\Gamma(1 + \nu + \frac{i\beta}{\gamma_+})\Gamma(1 - 2\nu)}.$$

From these relations we find that

$$A_- = \bar{A}_+, \quad B_- = \bar{B}_+, \quad C_- = \bar{C}_+, \quad D_- = \bar{D}_+ \quad .$$

Similar analysis as before shows that $n_- = 1$ for the parameter range $0 < \nu < \frac{1}{2}$ as well. Thus for the gapped graphene with a charge impurity, $n_+ = n_- = 1$ when $0 < \nu < \frac{1}{2}$. Therefore, this system admits a one parameter family of self-adjoint extensions for $0 < \nu < \frac{1}{2}$.

5.4.2 Scattering Matrix for Gapped Graphene with Coulomb Impurity

For the scattering states, we have $E > m$. In this case the parameter $\gamma = \sqrt{m^2 - E^2}$ becomes purely imaginary, i.e. $\gamma = iq$, where the real parameter q is defined as $q = \sqrt{E^2 - m^2}$. Consequently, the variable ρ also becomes purely imaginary, $\rho = 2iqr$.

The solutions which describes the scattering states have the form

$$P(\rho) = AM\left(\nu - \frac{\beta E}{\gamma}, 1 + 2\nu, \rho\right) + B\rho^{-2\nu}M\left(-\nu - \frac{\beta E}{\gamma}, 1 - 2\nu, \rho\right),$$

$$Q(\rho) = CM\left(1 + \nu - \frac{\beta E}{\gamma}, 1 + 2\nu, \rho\right) + D\rho^{-2\nu}M\left(1 - \nu - \frac{\beta E}{\gamma}, 1 - 2\nu, \rho\right).$$

With these solutions, we can write

$$\psi_1(r) = \sqrt{m + E}\; e^{-\frac{\rho}{2}}\rho^{\nu - \frac{1}{2}}\left[A(q)M\left(\nu - \frac{\beta E}{\gamma}, 1 + 2\nu, \rho\right)\right.$$

$$+ B(q)\rho^{-2\nu}M\left(-\nu - \frac{\beta E}{\gamma}, 1 - 2\nu, \rho\right)$$

$$+ C(q)M\left(1 + \nu - \frac{\beta E}{\gamma}, 1 + 2\nu, \rho\right)$$

$$\left. + D(q)\rho^{-2\nu}M\left(1 - \nu - \frac{\beta E}{\gamma}, 1 - 2\nu, \rho\right)\right],$$

where the coefficients $A(q)$ and $B(q)$ depend on the real parameter q. The functions P and Q are not independent of each other, but are related through the set of coupled equations

$$\rho\frac{dP}{d\rho} + (\nu - \frac{\beta E}{\gamma})P + \left(\frac{m\beta}{\gamma} - j\right)Q = 0,$$

$$\rho\frac{dQ}{d\rho} + (\nu - \rho + \frac{\beta E}{\gamma})Q - \left(j + \frac{m\beta}{\gamma}\right)P = 0.$$

As a consequence, the constants $A(q)$, $B(q), C(q)$ and $D(q)$ have relations between each other.

With this, we can express Q as

$$\left(j - \frac{m\beta}{\gamma}\right)Q = \left(\nu - \frac{\beta E}{\gamma}\right)\left[AM\left(\nu - \frac{\beta E}{\gamma}, 1 + 2\nu, \rho\right)\right.$$

$$\left. + B\rho^{-2\nu}M\left(-\nu - \frac{\beta E}{\gamma}, 1 - 2\nu, \rho\right)\right]$$

$$+ \rho\left[AM'\left(\nu - \frac{\beta E}{\gamma}, 1 + 2\nu, \rho\right) + B\rho^{-2\nu}M'\left(-\nu - \frac{\beta E}{\gamma}, 1 - 2\nu, \rho\right)\right.$$

$$\left. + B(-2\nu)\rho^{-2\nu-1}M\left(-\nu - \frac{\beta E}{\gamma}, 1 - 2\nu, \rho\right)\right]$$

where prime denotes the derivation with respect to ρ. Using the relation

$$aM(a, b, z) + zM'(a, b, z) = aM(a + 1, b, z)$$

(where prime denotes derivation with respect to z) we get

$$\left(j - \frac{m\beta}{\gamma}\right)Q = A(q)\left(\nu - \frac{\beta E}{\gamma}\right)M\left(1 + \nu - \frac{\beta E}{\gamma}, 1 + 2\nu, \rho\right)$$

$$+ B(q)\rho^{-2\nu}\left(-\nu - \frac{\beta E}{\gamma}\right)M\left(1 - \nu - \frac{\beta E}{\gamma}, 1 - 2\nu, \rho\right).$$

From this it follows that

$$C(q) = A(q)\frac{\nu - \frac{\beta E}{\gamma}}{j - \frac{m\beta}{\gamma}},$$

$$D(q) = B(q)\frac{-\nu - \frac{\beta E}{\gamma}}{j - \frac{m\beta}{\gamma}}.$$

As $r \rightarrow \infty$, and noting that $\gamma = iq$, the ψ_1 component of the wave function has the form

$$\psi_1(r) \longrightarrow \sqrt{m + E}\ (2iq)^{-\frac{1}{2} - i\frac{\beta E}{q}} \left[A(q)\frac{\Gamma(1 + 2\nu)}{\Gamma(1 + \nu - i\frac{\beta E}{q})} e^{-i\pi(\nu + i\frac{\beta E}{q})} \right.$$

$$\left. + B(q)\frac{\Gamma(1 - 2\nu)}{\Gamma(1 - \nu - i\frac{\beta E}{q})} e^{-i\pi(-\nu + i\frac{\beta E}{q})} \right] \frac{e^{-i(qr + \frac{\beta E}{q}\ln r)}}{\sqrt{r}}$$

$$+ \sqrt{m + E}\ (2iq)^{-\frac{1}{2} + i\frac{\beta E}{q}} \left[C(q)\frac{\Gamma(1 + 2\nu)}{\Gamma(1 + \nu + i\frac{\beta E}{q})} \right.$$

$$\left. + D(q)\frac{\Gamma(1 - 2\nu)}{\Gamma(1 - \nu + i\frac{\beta E}{q})} \right] \frac{e^{i(qr + \frac{\beta E}{q}\ln r)}}{\sqrt{r}}.$$

The scattering matrix can be obtained from the ratio of the outgoing and incoming amplitude, and can be written as

$$S(q) = e^{2i\delta(q)}$$

$$= (2iq)^{2i\frac{\beta E}{q}} \frac{C(q)\frac{\Gamma(1 + 2\nu)}{\Gamma(1 + \nu + i\frac{\beta E}{q})} + D(q)\frac{\Gamma(1 - 2\nu)}{\Gamma(1 - \nu + i\frac{\beta E}{q})}}{A(q)\frac{\Gamma(1 + 2\nu)}{\Gamma(1 + \nu - i\frac{\beta E}{q})} e^{-i\pi(\nu + i\frac{\beta E}{q})} + B(q)\frac{\Gamma(1 - 2\nu)}{\Gamma(1 - \nu - i\frac{\beta E}{q})} e^{-i\pi(-\nu + i\frac{\beta E}{q})}},$$

where δ denotes the phase shift.

In order to obtain the relations between the quantities A, B, C and D in the S-matrix, we need to impose the boundary conditions. As discussed before, the boundary conditions are encoded in the domain of self-adjointness. We have seen that the radial Dirac operator H_ρ and consequently Hamiltonian H admits a one parameter family of self-adjoint extension when $0 < \nu < \frac{1}{2}$. In this case, the domain of self-adjointness of the Hamiltonian is given by $\mathcal{D}_z(H) = \mathcal{D}(H) \oplus \{e^{i\frac{z}{2}}\Psi_+ + e^{-i\frac{z}{2}}\Psi_-\}$. In the limit $r \rightarrow 0$, a typical element of the domain $\mathcal{D}_z(H)$ can be written as

$$\Psi(\rho, \phi) = c\left(e^{i\frac{z}{2}}\Psi_+ + e^{-i\frac{z}{2}}\Psi_-\right),$$

where c is some constant and $\Psi_\pm$ are square integrable solutions of equations defining the deficiency indices. The $r \rightarrow 0$ limit of such a generic element of the domain $\mathcal{D}_z(H)$ can be obtained from the expression given in the previous section.

On the other hand, as $\rho \to 0$, , a component $\psi_1(\rho)$ of the physical wave function has the form

$$\psi_1(\rho) \longrightarrow \sqrt{m+E}\left((A(q)+C(q))\rho^{\nu-\frac{1}{2}} + (B(q)+D(q))\rho^{-\nu-\frac{1}{2}}\right).$$

Recall that $\rho = 2\gamma r$. By comparing the appropriate powers of r, we get

$$(2\gamma)^{\nu-\frac{1}{2}}\left(A(q)+C(q)\right)\sqrt{m+E} = ca\Big(\sqrt{m+i}e^{i\frac{z}{2}}(A_+ + C_+)$$
$$+ \sqrt{m-i}e^{-i\frac{z}{2}}(\bar{A}_+ + \bar{C}_+)\Big)(2\gamma_+)^{\nu-\frac{1}{2}}$$

$$(2\gamma)^{-\nu-\frac{1}{2}}\left(B(q)+D(q)\right)\sqrt{m+E} = -ca\Big(\sqrt{m+i}e^{i\frac{z}{2}}(B_+ + D_+)$$
$$+ \sqrt{m-i}e^{-i\frac{z}{2}}(\bar{B}_+ + \bar{D}_+)\Big)(2\gamma_+)^{-\nu-\frac{1}{2}},$$

where z is the self-adjoint extension parameter and all other quantities have been defined previously. The last two equations give

$$\frac{A(q)+C(q)}{B(q)+D(q)} = -\frac{\sqrt{m+i}e^{i\frac{z}{2}}(A_+ + C_+) + \sqrt{m-i}e^{-i\frac{z}{2}}(\bar{A}_+ + \bar{C}_+)}{\sqrt{m+i}e^{i\frac{z}{2}}(B_+ + D_+) + \sqrt{m-i}e^{-i\frac{z}{2}}(\bar{B}_+ + \bar{D}_+)}(2\gamma_+)^{2\nu}(2\gamma)^{-2\nu}$$
$$= -\frac{\xi_1 \cos(\theta_1 + \frac{z}{2})}{\xi_2 \cos(\theta_2 + \frac{z}{2})}(2\gamma_+)^{2\nu}(2\gamma)^{-2\nu},$$

where we have defined $\sqrt{m+i}(A_+ + C_+) \equiv \xi_1 e^{i\theta_1}$ and $\sqrt{m+i}(B_+ + D_+) \equiv \xi_2 e^{i\theta_2}$.

Using these equations, the S-matrix can be written as

$$S(q) =$$

$$(2iq)^{2i\frac{\beta E}{q}}\frac{\left[-\frac{\xi_1 \cos(\theta_1+\frac{z}{2})}{\xi_2 \cos(\theta_2+\frac{z}{2})}(2\gamma_+)^{2\nu}(2\gamma)^{-2\nu}\frac{1+\omega_2}{1+\omega_1}\omega_1\frac{\Gamma(1+2\nu)}{\Gamma(1+\nu+i\frac{\beta E}{q})} + \omega_2\frac{\Gamma(1-2\nu)}{\Gamma(1-\nu+i\frac{\beta E}{q})}\right]}{-\frac{\xi_1 \cos(\theta_1+\frac{z}{2})}{\xi_2 \cos(\theta_2+\frac{z}{2})}(2\gamma_+)^{2\nu}(2\gamma)^{-2\nu}\frac{1+\omega_2}{1+\omega_1}\frac{\Gamma(1+2\nu)}{\Gamma(1+\nu-i\frac{\beta E}{q})}e^{-i\pi(\nu+i\frac{\alpha E}{q})} + \frac{\Gamma(1-2\nu)}{\Gamma(1-\nu-i\frac{\beta E}{q})}e^{-i\pi(-\nu+i\frac{\beta E}{q})}},$$

where

$$\omega_1 \equiv \frac{C(q)}{A(q)} = \frac{\nu - \frac{\beta E}{\gamma}}{j - \frac{m\beta}{\gamma}}, \qquad \omega_2 \equiv \frac{D(q)}{B(q)} = \frac{-\nu - \frac{\beta E}{\gamma}}{j - \frac{m\beta}{\gamma}}.$$

This gives the S-matrix for gapped graphene for the parameter range $0 < \nu < \frac{1}{2}$. For this range of ν, the appropriate boundary conditions for which the Dirac Hamiltonian is self-adjoint and the corresponding time evolution unitary requires the introduction of an additional real self-adjoint extension parameter z, which labels the allowed boundary conditions. The phase shifts and the S-matrix depend explicitly on z. For each value of z (mod 2π), we have an inequivalent set of the scattering data. The above analysis by itself cannot determine which value of z would be realized in a given physical situation, which must be determined empirically.

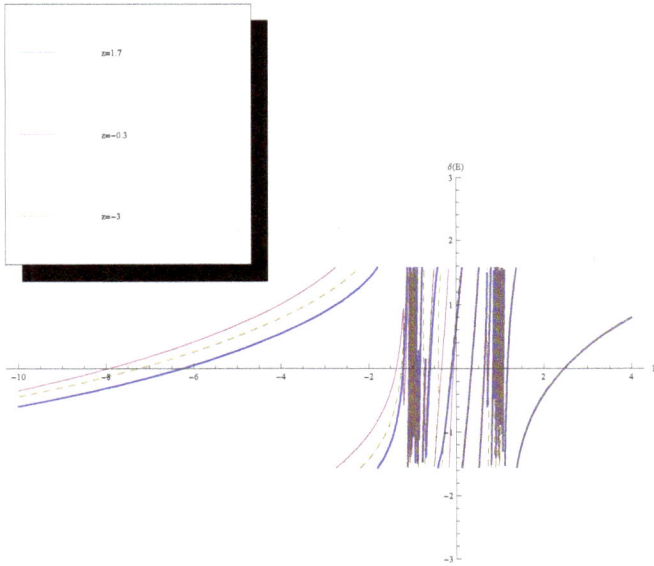

Fig. 5.6   Phase shifts in the gapped graphene for three different values of the self-adjoint extension parameter, $z = 1.7, \, -0.3, \, -3.0$ and for the values of the system parameters $j = \frac{3}{2}, \, \beta = 1.44, \, m = 1, \, \nu = 0.42$.

In Fig.5.6 we have plotted the scattering phase shift change with energy, for three different values of the self-adjoint extension parameter z. The phase shifts depend on the value of z, although its dependence on z is more marked for E, m. The oscillations in the energy region $-m < E < m$ indicate the appearance of discrete bound states.

Local density of states (LDOS) is defined by

$$\mu(E, r) = \frac{N_c}{\pi \hbar v_F} \sum_j |\Psi(r, \phi; E, j)|^2,$$

where N_c is the combined valley and spin degeneracy in graphene, which has the value $N_c = 4$. The LDOS has been calculated numerically and the dependence of the LDOS on the self-adjoint extension parameter z has been shown in Figs. 5.7 and 5.8.

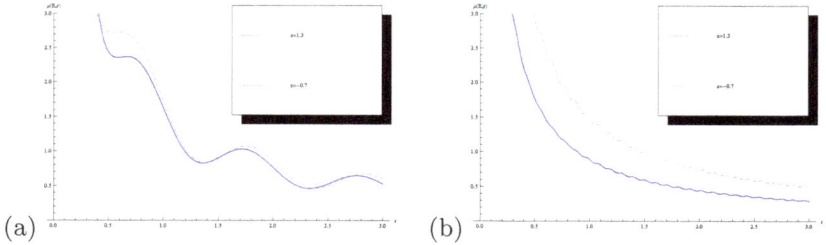

(a)   (b)

Fig. 5.7   (a) The local density of states in the scattering sector of the gapped graphene obtained by summing contributions coming from two angular momentum channels, $j = \frac{1}{2}$ and $j = -\frac{1}{2}$, where both contributions correspond to energy $E = 3$. Local densities are drawn for two different values $z = 1.3$, -0.7 and for the values of the system parameters $\alpha = 0.49$, $m = 1$, $\nu = 0.1$ and $N_c = 4$. (b) Local density of states in the scattering sector for the same parameters except it corresponds to energy $E = 30$.

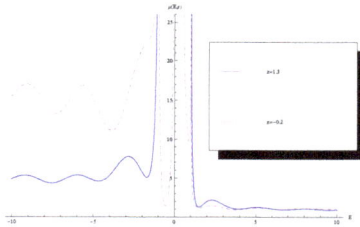

Fig. 5.8   Energy dependence of the local density of states in the gapped graphene at the distance $r = 1$ from the impurity, obtained by summing the contributions coming from the angular momentum channels $j = \frac{1}{2}$ and $j = -\frac{1}{2}$. Local densities are drawn for two different values of the parameter z, $z = 1.3$, -0.2 and for the values of the system parameters $\alpha = 0.49$, $m = 1$, $\nu = 0.1$ and $N_c = 4$.

Figures 5.7(a) and 5.7(b) show LDOS in the scattering sector of the gapped graphene as a function of the distance from the impurity. LDOS depicted at Fig.5.7(a) corresponds to situation when energy is kept fixed at $E = 3m$, while Fig.5.7(b) corresponds to energy ten times greater, $E = 30m$. Figure 5.8 shows the energy dependence of LDOS in gapped graphene, at a location close to the charge impurity.

The bound states correspond to the poles of the S matrix in the positive

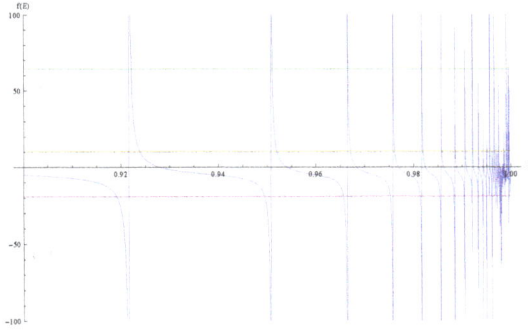

Fig. 5.9   Graph of the left hand side of energy equation for gapped graphene with $j = \frac{3}{2}$, $\alpha = 1.44$, $m = 1$ and $\nu = 0.42$. Horizontal straight lines correspond to values of the right hand side of energy equation for $z = 1.0$, -0.7, 1.9 from top to bottom.

imaginary axis. The bound state energies are given by the solutions of the equation

$$
\left(\frac{m^2 - E^2}{m^2 + 1} \right)^{\nu} \frac{\Gamma(1 - 2\nu)\Gamma(1 + \nu - \frac{\beta E}{\sqrt{m^2 - E^2}})}{\Gamma(1 + 2\nu)\Gamma(1 - \nu - \frac{\beta E}{\sqrt{m^2 - E^2}})} \left(\frac{j + \nu + \frac{\beta(m+E)}{\sqrt{m^2 - E^2}}}{j - \nu + \frac{\beta(m+E)}{\sqrt{m^2 - E^2}}} \right)
$$
$$
= \frac{\xi_1 \cos(\theta_1 + \frac{z}{2})}{\xi_2 \cos(\theta_2 + \frac{z}{2})}.
$$

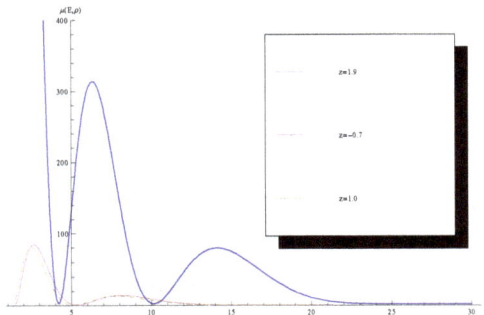

Fig. 5.10   Contribution to the local density of states in the bound state sector of the gapped graphene, coming from the angular momentum channel $j = \frac{3}{2}$. This contribution is calculated for the values of the system parameters $\alpha = 1.44$, $m = 1$, $\nu = 0.42$ and for three different values of $z = 1.0$, -0.7, 1.9.

The solutions of this equation gives the bound state energies in gapped graphene in the presence of a charge impurity, for the parameter range $0 < \nu < \frac{1}{2}$. This is solved numerically in Fig. 5.9. For a given set of system parameters, each horizontal line in Fig. 5.9 corresponds to a particular value of z. The plot shows that the spectrum explicitly depends on the choice of z. This is not surprising since z labels all the allowed boundary conditions and the plot in Fig. 5.9 depicts the effect of the boundary condition on the spectrum. Fig. 5.10 plots the LDOS coming from the bound state sector for three different values of the self-adjoint extension parameter z.

5.5   Gapped   Graphene   with   a   Supercritical   Coulomb   Charge

The supercritical regime sets in when the effective Coulomb strength $\beta^2 > j^2$ for any angular momentum channel j. This implies that in the supercritical regime $\beta > \frac{1}{2}$ and $\nu = \sqrt{j^2 - \beta^2} = \pm i\mu$ where $\mu \in R$. We shall study the Dirac equation in this regime. We focus on the excitations satisfying $E^2 < m^2$ and introduce a cutoff in the radial direction set by the lattice spacing in graphene. The cutoff restricts our analysis to the region where the Dirac equation holds. The corresponding eigenvalue problem is solved with a hard-core boundary condition given by

$$\psi(\rho = \rho_0) = 0, \quad \rho_0 = 2r_0\gamma,$$

where ψ is the two component wavefunction and r_0 provides the cutoff in the radial direction. In this case, the upper component $\psi_1(\rho)$ has two linearly independent solutions given by

$$\xi(\rho) = \sqrt{m + E}\, e^{-\frac{\rho}{2}} \rho^{i\mu - \frac{1}{2}} M\left(i\mu - \frac{\beta E}{\gamma}, 1 + 2i\mu, \rho\right)$$

$$\zeta(\rho) = \sqrt{m + E}\, e^{-\frac{\rho}{2}} \rho^{-i\mu - \frac{1}{2}} M\left(-i\mu - \frac{\beta E}{\gamma}, 1 - 2i\mu, \rho\right).$$

The general solution which satisfies the above boundary condition can be written as

$$\psi_1(\rho) = [\xi(\rho)\zeta(\rho_0) - \zeta(\rho)\xi(\rho_0)].$$

As $\rho \longrightarrow \infty$, we find

$$\psi_1(\rho) \longrightarrow \sqrt{m + E}\, e^{\frac{\rho}{2}} \rho^{-\frac{1}{2}} \left[\frac{\Gamma(1 + 2i\mu)}{\Gamma(i\mu - \frac{\alpha E}{\gamma})} \zeta(\rho_0) - \frac{\Gamma(1 - 2i\mu)}{\Gamma(-i\mu - \frac{\alpha E}{\gamma})} \xi(\rho_0) \right].$$

In order for the wave function to be square integrable, the quantity in the parenthesis which multiplies $e^{\frac{\rho}{2}}$ must vanish. This gives the condition

$$\frac{\Gamma(1+2i\mu)\Gamma(-i\mu-\frac{\alpha E}{\gamma})}{\Gamma(1-2i\mu)\Gamma(i\mu-\frac{\alpha E}{\gamma})} = \frac{\xi(\rho_0)}{\zeta(\rho_0)}.$$

The above condition is exact but it is not easy to proceed further with it. In order to gain some physical insight, we shall now use several approximations. The results discussed below are therefore valid only in a qualitative fashion. First we assume that as the cutoff r_0 approaches the lattice spacing, the hypergeometric function M can be replaced approximately by 1. Strictly speaking this is true when the cutoff tends to zero, but it is a reasonable approximation in the long wavelength limit. Second, we assume that $E^2 < m^2$. In other words, we shall trust our results only for energy scales below the Dirac mass. Using these assumptions, we get

$$\frac{\Gamma(-i\mu-\frac{\beta E}{\gamma})}{\Gamma(i\mu-\frac{\beta E}{\gamma})} = e^{2i(\mu\ln\rho_0+\delta)},$$

where δ is the argument of $\Gamma(1-2i\mu)$. In order to proceed, consider the energy scale such that $\frac{E}{m} << 1$. In this case, the left hand side of the above equation is approximately independent of E and depends only on the system parameter μ. Denoting the argument of $\Gamma(-i\mu)$ by θ, we get

$$\gamma_p = \frac{1}{2r_0}e^{\frac{\theta-\delta-2p\pi}{\mu}},$$

where $p \in Z$ and $\gamma_p = \sqrt{m^2 - E_p^2}$. We can satisfy the requirement of $\frac{E}{m} << 1$ by restricting p suitably. We now assume that μ, through its dependence on the effective Coulomb coupling β, is a function of the cutoff r_0. We keep E_p or equivalently γ_p invariant as the cutoff is varied, which gives the beta function $\tilde{\beta}$ function as

$$\tilde{\beta}(\mu) = -r_0\frac{d\mu}{dr_0} \sim -\mu^2.$$

We see that the beta function admits an ultraviolet stable fixed point at $\mu = 0$ or equivalently at $\alpha = j$ for the angular momentum channel j, to which the system is expected to flow. In particular, the Coulomb coupling tends to its critical value $\frac{1}{2}$ for the angular momentum channel $j = \frac{1}{2}$.

5.6 Graphene with Charge Impurity and Topological Defects

Topological defects are of great interest in quantum theories. Defects such as strings, monopoles and textures gives rise to fascinating predictions, some of which are not so easy to observe in the laboratory. Consider for example the dynamics of Dirac fermions in the presence of a cosmic string. Such a system is of great interest in studying quantum aspects of gravity. It has been predicted some time back that in the presence of cosmic strings, the Dirac fermions exhibit unusual quantum properties. In particular, the quantum theory for such a system is not unique, and it admits a one-parameter family of inequivalent quantizations. It is however very hard to imagine a laboratory setup where such a prediction of gravity could be tested. The discovery of graphene now provides an opportunity to test such predictions empirically.

As discussed before, the low energy excitations in graphene obey a 2D Dirac equation. If the sublattice symmetry is violated, the Dirac fermions would be massive and the system would be gapped. When a conical topological defect is introduced in such a system, it gives rise to nontrivial holonomies. The boundary conditions associated with the holonomies can be realized by introducing a suitable flux tube, analogous to a cosmic string. We shall use such a flux tube to model the conical topological defect on the 2D graphene sheet. In addition to the topological defect, we shall also include a Coulomb charge impurity in the system. Previously we have studied the effect of the charge impurity alone in graphene. Now we shall see how the topological defect modifies the effect of the charge impurity. As discussed before, the charge impurity strength could be either sub or super critical. The value $\beta_c = \frac{1}{2}$ was the minimum possible value of the critical charge for a graphene plane. We shall show below that a string type topological defect has the possibility to modify the critical charge in graphene. This discussion would be done in the context of gapped graphene.

Recall that graphene has a hexagonal honeycomb lattice structure which is formed by two interpenetrating triangular sublattices A and B. The low energy excitations have minimum energy eigenvalues at the six vertices of the Brillouin zone, which are called the Dirac points. Among these only two are inequivalent. We denote their wave vectors by $\mathbf{K}_+$ and $\mathbf{K}_-$. The low-energy properties of the quasiparticle states in graphene near the Dirac point K_+ can be described by the Dirac equation

$$H\Psi = [-i(\sigma_1\partial_x + \sigma_2\partial_y) + m\sigma_3]\Psi = E\Psi,$$

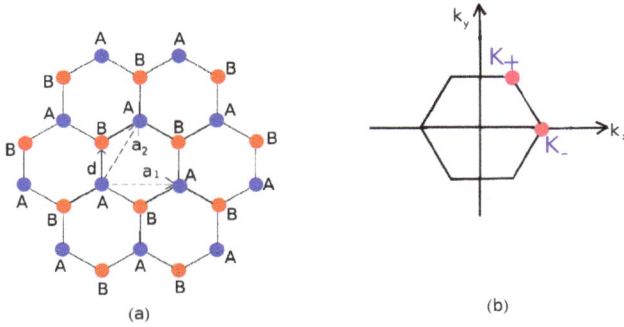

Fig. 5.11   Hexagonal lattice structure of graphene and the Brillouin zone.

where m denotes the Dirac mass, E is the energy eigenvalue and we have set $\hbar = v_F = 1$. For this discussion the mass m will play no special role and it has been included for generality. The Dirac Hamiltonian, which is valid at the long wavelength limit, acts on slowly varying wavefunction $\Psi = \begin{pmatrix} \Psi_{K_+,A} \\ \Psi_{K_+,B} \end{pmatrix}$ and the Pauli matrices $\sigma_{1,2,3}$ act on the pseudospin indices A, B.

To study the effect of topology on this system, we introduce a conical defect in the system. As shown in Fig. 5.12, this is achieved by removing a wedge OAB and then joining the two line segments OA and OB. From this diagram it is clear that the number n of such segments that can be removed goes from 1 to 5. $n = 0$ corresponds to planar graphene without any topological defect. For a given value of n, the deficit angle is given by $\frac{2n\pi}{6}$.

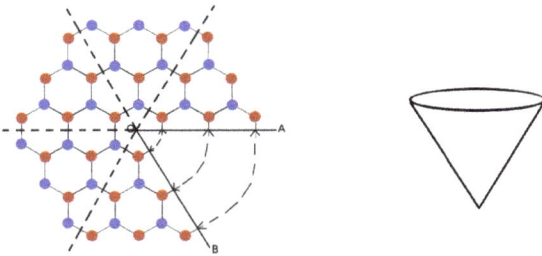

Fig. 5.12   Formation of a cone from planar graphene.

In a plane, the low energy quasiparticles of graphene behave as electrons. A conical defect results in nontrivial holonomies for the quasiparticle wavefunctions. If a quasiparticle is taken around a loop enclosing the defect, the angular part of the corresponding wavefunction obeys the condition given by

$$\Psi(r, \theta = 2\pi) = e^{i2\pi(1-\frac{n}{6})\frac{\sigma_3}{2}}\Psi(r, \theta = 0),$$

where (r, θ) denotes the polar coordinates on the plane.

When the cone is formed by removing an odd number of wedges of angle $\frac{2\pi}{6}$ an additional phase shift appears affecting the valley indices of the wave function in the boundary condition. The states with valley index K_- will be affected in the same manner as the states with valley index K_+ but there will be a relative phase difference of $180°$ between them. Therefore this boundary condition can be described by involving a τ_2 matrix, which operates on the valley indices. When n is even, this off diagonal matrix does not play any role and the exponential factor appearing in the boundary condition just gives ± 1 depending on the value of n. We diagonalize the matrix τ_2 for all allowed odd values of n. As a result the valley indices of the electronic states become mixtures of K_+ and K_-. Then the angular boundary condition satisfied for all values of n is given by

$$\Psi(r, \theta = 2\pi) = e^{i2\pi[\pm\frac{n}{4}\sigma_0 + (1-\frac{n}{6})\frac{\sigma_3}{2}]}\Psi(r, \theta = 0).$$

Here σ_0 is an identity matrix which acts on the pseudospin indices A, B. $\Psi = \begin{pmatrix} \Psi_{A,K'} \\ \Psi_{B,K'} \end{pmatrix}$ where K' is a mixture of K_+ and K_-.

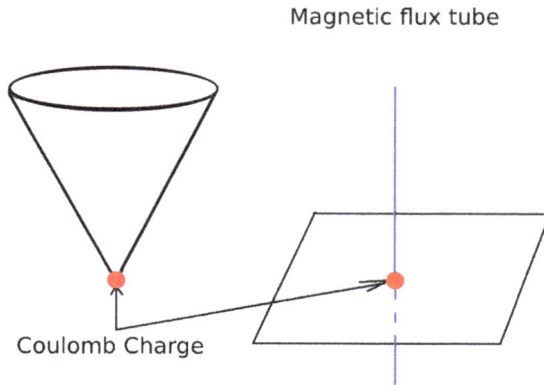

Fig. 5.13   Graphene cone = planar graphene + flux tube.

The effect of the holonomy can be modelled by introducing a fictitious magnetic flux tube passing through the apex of the cone. The magnetic vector potential modifies the boundary condition on a Dirac spinor as

$$\Psi(r, \theta = 2\pi) = e^{ie \oint \vec{A} \cdot \vec{ds}} \Psi(r, \theta = 0).$$

Here $\vec{ds}$ is a line element on the circumference of the cone at a distance r from the apex,

$$\vec{ds} = \hat{e}_\theta \, r(1 - \frac{n}{6})d\theta.$$

Using these equations we can get an expression of the fictitious vector potential as

$$A_\theta = \frac{1}{er}[\pm \frac{\frac{n}{4}\sigma_0}{(1 - \frac{n}{6})} + \frac{\sigma_3}{2}].$$

Thus an external Coulomb charge localized at the apex of the gapped graphene cone can be equivalently described by a suitable combination of electric charge and magnetic flux tube.

The corresponding Dirac equation in the presence of the external Coulomb charge and the topological defect can be written as

$$H\Psi(r, \theta) = E\Psi(r, \theta), \tag{5.1}$$

where

$$H = \begin{pmatrix} m - \frac{\beta}{r} & \partial_r - \frac{i}{r(1-\frac{n}{6})}\partial_\theta \pm \frac{\frac{n}{4}}{r(1-\frac{n}{6})} + \frac{1}{2r} \\ -\partial_r - \frac{i}{r(1-\frac{n}{6})}\partial_\theta \pm \frac{\frac{n}{4}}{r(1-\frac{n}{6})} - \frac{1}{2r} & -m - \frac{\beta}{r} \end{pmatrix}$$

and the wavefunction $\Psi(r, \theta)$ can be expressed as

$$\Psi(r, \theta) = \begin{pmatrix} \Psi_A(r, \theta) \\ \Psi_B(r, \theta) \end{pmatrix} = \sum_j \begin{pmatrix} \Psi_A^{(j)}(r) \\ \Psi_B^{(j)}(r) \end{pmatrix} e^{ij\theta},$$

where j is half-integer.

Substituting the wavefunction in the Dirac equation we find that the short distance nature of the radial part of the wavefunction has the form

$$\Psi_{A,B}^{(j)}(r) \sim r^{\sqrt{\gamma^2 - \beta^2} - \frac{1}{2}} \quad \text{and} \quad \gamma = \frac{(j \pm \frac{n}{4})}{(1 - \frac{n}{6})}.$$

This analysis already provides a very important information about the effect of topology on this system. To see that, first note that in the planar case, where $n = 0$, the short distance behaviour of the wavefunction was given by $r^{\sqrt{j^2 - \beta^2} - \frac{1}{2}}$. Now j has been replaced by γ which is a function of both j and the number of sectors n removed from the graphene plane to form the cone. In the presence of the topological defect, if the strength of the external charge impurity characterized by β exceeds that value of γ for any value of j and n, the wavefunction would show wild oscillatory behaviour near the origin. As discussed, such a behaviour indicates the onset of quantum instability due to strong nonperterbative QED effects. In the planar case, the critical charge was defined by $\beta_c = \frac{1}{2}$. What is most interesting is that in the presence of the topological defect, the value of the critical charge is determined not just by j but by γ which depends on both j and n. The topological defect has a direct effect on the onset of the strong QED regime, which illustrates a beautiful interplay between topology and dynamics.

A particularly interesting situation occurs in the angular momentum channel $j = \frac{1}{2}$. There, for a suitable topology, namely for $n = 2$ and for the choice of the negative sign in the numerator of γ, we see that $\gamma = 0$. This means that the critical charge vanishes. In other words, for $j = \frac{1}{2}$ and for $n = 2$, any external charge in the system, no matter how small, would be supercritical and would lead to quantum instability. This is a very robust prediction of this analysis which should be amenable to experimental verification.

Further Reading and Selected References

The following papers provide good reviews on various electronic properties of graphene : A. H. Castro Neto et al, Rev. Mod. Phys. **81**, 109 (2009); V. N. Kotov et al, Rev. Mod. Phys. **84**, 1067 (2012).

For Klein tunneling in graphene see M. I. Katsnelson et al, Nature Physics **2**, 620 - 625 (2006).

For discussions on supercritical charges in graphene see V. M. Pereira et al, Phys. Rev. Lett. **99**, 166802 (2007); A. V. Shytov et al, Phys. Rev. Lett. **99**, 246802 (2007); Y. Wang et al, Science, **340**, 734 (2013). The review by N. Kotov et al mentioned above also has a discussion on supercritical regime of graphene.

There are several papers on topological defects in graphene. Some of them are P.E. Lammert, V.H. Crespi, Phys. Rev. Lett. **85**, 5190 (2000); P.E. Lammert, V.H. Crespi, Phys. Rev. B **69**, 035406 (2004); A. Cortijo, M.A.H. Vozmediano, Nucl. Phys. **B 763**, 293 (2007); B. Chakraborty et al, Phys. Rev. **B 83** (2011) 115412; B. Chakraborty et al, J. Phys. A46 (2013) 055303; B. Chakraborty et al, J. Phys. Conf. Ser. 442 (2013) 012017; J. K. Pachos and Michael Stone, Int. J. Mod. Phys. **B21** (2007) 5113-5120; Abhishek Roy and Michael Stone, J. Phys. **A43** (2010) 015203.

For application of self-adjoint extension in graphene see Kumar S. Gupta and S. Sen, Mod. Phys. Lett. **A24** (2009) 99; Kumar S. Gupta and S. Sen, Phys. Rev. **B 78** (2008) 205429; Kumar S. Gupta et al, Eur. Phys. J. **B 73** (2010) 389. Figures 5.2–5.10 are reprinted from the latter with permission of Springer Science & Business Media.

For a connection of supercritical regime of graphene with Riemann zeta function see Kumar S. Gupta et al, EPL **102** (2013) 10006.

Index